AF830090

Emission Trajectories and Mitigation Schemes for China

This book seeks to model the possible emission trajectories and identify the feasible mitigation schemes for China to meet its climate commitments to peak emissions before 2030 and net zero emissions before 2060.

In line with these ambitions, China has taken a number of measures to improve carbon efficiency and energy structure in recent years. The book first analyzes changes in the carbon footprint at the city level, identify the different pathways to peak emissions by province and industry, and develop a bottom-up approach to determining when and how China could reach peak emissions. It then illustrates how the Emissions Trading Scheme (ETS) reduces abatement costs, and examines the cost-saving effects under carbon intensity reduction and peak emissions targets. Based on the findings and the status quo in China, the authors propose a multi-criteria allocation scheme for carbon emission allowances at the provincial level and quantify the impact of sectoral coverage on the abatement costs of the ETS by developing a cost-based approach to sectoral coverage in China. In addition, the co-benefits between CO_2 and $PM_{2.5}$ reductions as a result of the ETS operation are elaborated.

The book will benefit researchers and policy-makers interested in environmental governance, climate policy, environmental economics, and sustainable development.

Kai Fang is a Full Professor of Environmental Management at the School of Public Affairs, Zhejiang University, China. His research interests include environmental and climate policy and sustainable development.

Qifeng Zhang is a Policy Researcher at Huaneng Yangtze Environmental Technology Research Institute, Beijing, China. His research covers environmental and economic policy, and resource and environmental management.

Yiqi Tang is a PhD candidate at the School of Public Policy and Management, Tsinghua University, China. Her research covers environmental and climate policy and public governance in transitional China.

Emission Trajectories and Mitigation Schemes for China

Kai Fang, Qifeng Zhang, and Yiqi Tang

Routledge
Taylor & Francis Group
LONDON AND NEW YORK

First published in English 2024
by Routledge
4 Park Square, Milton Park, Abingdon, Oxon OX14 4RN

and by Routledge
605 Third Avenue, New York, NY 10158

Routledge is an imprint of the Taylor & Francis Group, an informa business

British Library Cataloguing-in-Publication Data
A catalogue record for this book is available from the British Library

Library of Congress Cataloging-in-Publication Data
Names: Fang, Kai, 1986- author. | Zhang, Qifeng, 1994- author. | Tang, Yiqi, 1997- author.
Title: Emission trajectories and mitigation schemes for China / Kai Fang, Qifeng Zhang, Yiqi Tang.
Description: Abingdon, Oxon ; New York, NY : Routledge, 2024. | Originally published in Chinese by China Renmin University Press. | Includes bibliographical references and index. |
Identifiers: LCCN 2023022235 (print) | LCCN 2023022236 (ebook) | ISBN 9781032560953 (hardback) | ISBN 9781032560977 (paperback) | ISBN 9781003433859 (ebook)
Subjects: LCSH: Carbon offsetting--China. | Carbon dioxide mitigation--China. | Environmental policy--Ghina.
Classification: LCC HC430.P55 F36 2024 (print) | LCC HC430.P55 (ebook) | DDC 363.73/70951--dc23/eng/20230706
LC record available at https://lccn.loc.gov/2023022235
LC ebook record available at https://lccn.loc.gov/2023022236

ISBN: 978-1-032-56095-3 (hbk)
ISBN: 978-1-032-56097-7 (pbk)
ISBN: 978-1-003-43385-9 (ebk)

DOI: 10.4324/9781003433859

Typeset in Times New Roman
by KnowledgeWorks Global Ltd.

Contents

Figures

Tables

Acknowledgements

This work was supported by the Major Project of National Social Science Fund of China (22&ZD108), and the National Natural Science Foundation of China (72074193, 42341205).

1 Introduction

Numerous evidence has demonstrated that global warming caused by anthropogenic greenhouse gases (GHG) emissions has posed unprecedented threats to the international community, such as extreme weather, sea-level rise, infectious diseases, biodiversity loss, and food shortage (Cavicchioli et al., 2019; Le Quéré et al., 2019; Martinich & Crimmins, 2019; Raza et al., 2019). Economic growth has been leading to a continuous increase in global emissions, reaching a record-high level of 33.3 Gt in 2019 (Friedlingstein et al., 2020 & 2022; IEA, 2020). While this trend has been reversed during the COVID-19 pandemic, with a 17% reduction in global daily emissions, compared with the average level in 2019, this reduction may be temporary, because no evidence supports any structural changes to the world's industrial systems (Chong et al., 2022; Le Quéré et al., 2020; Tollefson, 2021). Retaliatory emissions are likely to occur after production and consumption activities recover. To mitigate climate change, a consensus has been reached worldwide that the end-of-century warming must be maintained below 2°C, or even 1.5°C above the pre-industrial levels (IPCC, 2022; Tong et al., 2019). In response to this target, over 190 nations and regions have pledged their nationally determined contributions (NDCs) to reduce GHG emissions and become carbon-neutral in the next coming decades (Glanemann et al., 2020; Weikmans et al., 2020).

As the world's largest GHG emitter since 2006, China has proposed numerous climate targets aimed at decoupling economic growth from emissions, reducing emissions per unit of GDP (i.e., carbon intensity) and peaking the emissions. For example, China's National Climate Change Program published in 2007 pledged 20% reduction of energy consumption per unit of GDP (i.e., energy intensity) by 2010 for consequently reducing carbon emissions. In addition to the controls on energy consumption and intensity, China first committed to reducing carbon intensity by 40%–45% by 2020 against the 2005 levels in 2009 (Zhang et al., 2009, 2019). Furthermore, such a carbon-oriented goal has been officially included in China's 12th Five Year Plan (2011–2015) in which 17% reduction of carbon intensity has to be achieved (Zheng et al., 2015). These climate commitments were further updated as NDCs in which the carbon intensity is reduced by 60%–65% in 2030 above the 2005 levels, and that peaking emissions are achieved by latest 2030. In 2021, China submitted its strengthened version of NDCs that commits to reducing the carbon intensity by over 65% in 2030 against the 2005 levels, as well

DOI: 10.4324/9781003433859-1

as to reaching carbon peaking by 2030 and carbon neutrality by 2060 (Fang et al., 2021; Robinson et al., 2021).

In striving to fulfill these ambitions, China has taken a basket of measures to enhance energy efficiency and improve energy mix over past years. Circular economy strategy at the city level is one example that redefines products and services while minimizing negative environmental impacts (Dong et al., 2016; Mathews & Tan, 2016). Various related approaches and policies are promoted, including but not limited to cleaner production, 3R (i.e., reduce, reuse and recycle) promotion, eco-industrial parks, industrial symbiosis, eco-town and zero-emission practices, and Sustainable Development Goals (SDGs) (Geng & Doberstein, 2008; Geng et al., 2010; He et al., 2022). The evaluation on these approaches and policies are inevitably indispensable not only for investigating their contributions to emission reduction, but also for adapting the policies to China's dual carbon goals towards 2030 and 2060, respectively.

Furthermore, additional to the critical role of these city-wise policies in reducing carbon emissions, there is a need for investigation into whether and when China will peak its emissions at multiple scales (Fang et al., 2019 & 2022; Shan et al., 2022). Because of the heterogeneity of socioeconomic and environmental dimensions across regions and industrial sectors, a multi-provincial and multi-sectoral perspective that suffices to provide novel insights into the potential peak emission pathways in China is highly required (Green & Stern, 2015; Niu et al., 2016). Yet, to our knowledge, future emission pathways for various provinces and industries throughout China have not been adequately explored in the existing literature.

Responding to China's climate goals, market-based solutions have come into force in a cost-effective manner. Emission trading scheme (ETS) that makes use of carbon emission allowance (CEA) or reduction quotas trading is one example that plays a central role in meeting the NDCs and saving abatement costs for numerous regions and countries. Specifically, European Union (EU) pioneered to operating the first multinational ETS of the world in 2005 in response to the Kyoto Protocol and has witnessed the rapid progress since then (Convery, 2009; Flachsland et al., 2020). However, despite the dramatic development during the first two commitment periods (i.e., 2005–2007, and 2008–2012), EU ETS has faced the great recession since 2013 when the carbon price dropped to €2.75 and remained at this low level for five years. In case of little incentives to invest in low-carbon technologies under such a low carbon price, a series of reforms have been adopted since 2018, including but not limited to tightening the emission cap, modifying the market stability reserve (MSR) (Brink et al., 2016; Löfgren et al., 2014; Sanin et al., 2015). Literally, the therapies that brought the carbon price back to levels around 25–30€/t and made the ETS come into effect provide insightful references for this market paradigm in other countries or regions (Perino, 2018; Zhang & Wei, 2010).

Likewise, ETS has been regarded as the most effective market-based tool in mitigating the climate change in China (Jotzo & Löschel, 2014; Zhang, 2015; Zhang et al., 2014). Since the first ETS pilot initiated in Shenzhen in 2013, China has implemented the ETS in eight pilot provinces and cities, e.g., Beijing, Shanghai,

Tianjin. The lessons derived from these pilots accelerates the implementation of China's nation-wide ETS. In 2021, China officially launched the nation-wide ETS that first covers the power sector accounting for over 40% of the total carbon emissions. It is reported that more than 50 million tons of CEA are traded via China's nation-wide ETS at price of 50.54–61.60 RMB per ton in 2022. However, despite the great success of the nation-wide ETS, a series of challenges remain that not only impede the development of China's ETS but also distort the cost-effectiveness in analogy to the EU ETS, e.g., inactive trading, low carbon price, disincentive effects on low-carbon investment (Liu et al., 2015; Lo, 2016; Tang et al., 2018). There is a lack of consensus on setting emission cap, determining sectoral coverage, and allocating CEA. Referring to the lessons learned from the EU ETS, bringing clarity to these key questions becomes a top priority for running the China's ETS in a cost-effective manner.

In keeping with China's commitments towards peak emissions in 2030 and net-zero emissions in 2060, primarily this book is intended to model the possible emission trajectories and identify the feasible mitigation schemes for China. As such, this book is structured as follows. In response to China's dual carbon goals, Chapters 2–4 generally conduct the ex-ante exploration on whether and how China could peak its emissions and become carbon neutral at different levels. Specifically, Chapter 2 sheds lights on the city level, traces the evolution of their carbon footprint and investigates the role of a city in combating climate change. Chapters 3 and 4 focus on the industries and provinces and identify their different pathways towards peak emissions by developing a bottom-up approach to exploring when and how China could reach peak emissions. In addition to these ex-ante explorations, Chapter 5 takes a further look at the cost-effective manners in mitigating climate change (i.e., the nation-wide ETS), and illustrates how the ETS reduces the abatement costs and examines the cost-saving effects. Concerning the challenges faced by China's nation-wide ETS, Chapters 6 and 7 are intended to bring more clarity to the policy design and reforms. Of which, Chapter 6 proposes a multi-criteria allocation scheme for CEA at the provincial level, and Chapter 7 quantifies the impacts of sectoral coverage on ETS's abatement costs by developing a costs-oriented approach for sectoral coverage (COASCO) in China. Chapter 8 explores the evaluation of co-benefits between CO_2 and $PM_{2.5}$ reduction as a result of the ETS operation.

References

Brink, C., Vollebergh, H. R., & van der Werf, E. (2016). Carbon pricing in the EU: Evaluation of different EU ETS reform options. *Energy Policy*, *97*, 603–617.

Cavicchioli, R., Ripple, W. J., Timmis, K. N., Azam, F., Bakken, L. R., Baylis, M., & Webster, N. S. (2019). Scientists' warning to humanity: Microorganisms and climate change. *Nature Reviews Microbiology*, *17*(9), 569–586.

Chong, C. T., Van Fan, Y., Lee, C. T., & Klemeš, J. J. (2022). Post COVID-19 ENERGY sustainability and carbon emissions neutrality. *Energy*, *241*, 122801.

Convery, F. J. (2009). Origins and development of the EU ETS. *Environmental and Resource Economics*, *43*(3), 391–412.

Dong, L., Fujita, T., Dai, M., Geng, Y., Ren, J., Fujii, M., Wang, Y., & Ohnishi, S. (2016). Towards preventative eco-industrial development: An industrial and urban symbiosis case in one typical industrial City in China. *Journal of Cleaner Production*, *114*, 387–400.

Fang, K., Li, C., Tang, Y., He, J., & Song, J. (2022). China's pathways to peak carbon emissions: New insights from various industrial sectors. *Applied Energy*, *306*, 118039.

Fang, K., Tang, Y., Zhang, Q., Song, J., Wen, Q., Sun, H., & Xu, A. (2019). Will China peak its energy-related carbon emissions by 2030? Lessons from 30 Chinese provinces. *Applied Energy*, *255*, 113852.

Fang, K., Zhang, Q., Song, J., Yu, C., Zhang, H., & Liu, H. (2021). How can national ETS affect carbon emissions and abatement costs? Evidence from the dual goals proposed by China's NDCs. *Resources, Conservation and Recycling*, *171*, 105638.

Flachsland, C., Pahle, M., Burtraw, D., Edenhofer, O., Elkerbout, M., Fischer, C., & Zetterberg, L. (2020). How to avoid history repeating itself: The case for an EU emissions trading system (EU ETS) price floor revisited. *Climate Policy*, *20*(1), 133–142.

Friedlingstein, P., Jones, M. W., O'Sullivan, M., Andrew, R. M., Bakker, D. C., Hauck, J., & Zeng, J. (2022). Global carbon budget 2021. *Earth System Science Data*, *14*(4), 1917–2005.

Friedlingstein, P., O'sullivan, M., Jones, M. W., Andrew, R. M., Hauck, J., Olsen, A., & Zaehle, S. (2020). Global carbon budget 2020. *Earth System Science Data*, *12*(4), 3269–3340.

Geng, Y., & Doberstein, B. (2008). Developing the circular economy in China: Challenges and opportunities for achieving 'leapfrog development. *The International Journal of Sustainable Development & World Ecology*, *15*(3), 231–239.

Geng, Y., Tsuyoshi, F., & Chen, X. (2010). Evaluation of innovative municipal solid waste management through urban symbiosis: A case study of Kawasaki. *Journal of Cleaner Production*, *18*, 993–1000.

Glanemann, N., Willner, S. N., & Levermann, A. (2020). Paris climate agreement passes the cost-benefit test. *Nature Communications*, *11*(1), 1–11.

Green, F., & Stern, N. (2015). China's 'new normal': Structural change, better growth, and peak emissions. Centre for Climate Change Economics and Policy, 7–10.

He, J., Yang, Y., Liao, Z., Xu, A., & Fang, K. (2022). Linking SDG 7 to assess the renewable energy footprint of nations by 2030. *Applied Energy*, *317*, 119167.

IEA. (2020). Global CO_2 emissions in 2019. https://www.iea.org/articles/global-co2-emissions-in-2019

IPCC. (2022). *Climate change 2022: Impacts, adaptation, and vulnerability. Contribution of working group II to the sixth assessment report of the intergovernmental panel on climate change.* Intergovernmental Panel on Climate Change (IPCC).

Jotzo, F., & Löschel, A. (2014). Emissions trading in China: Emerging experiences and international lessons. *Energy Policy*, *75*, 3–8.

Le Quéré, C., Jackson, R. B., Jones, M. W., Smith, A. J. P., Abernethy, S., & Andrew, R. M., et al. (2020). Temporary reduction in daily global CO_2 emissions during the COVID-19 forced confinement. *Nature Climate Change*, *10*(7), 1–7.

Le Quéré, C., Korsbakken, J. I., Wilson, C., Tosun, J., Andrew, R., Andres, R. J., Canadell, J. G., Jordan, A., Peters, G. P., & van Vuuren, D. P. (2019). Drivers of declining CO_2 emissions in 18 developed economies. *Nature Climate Change*, *9*, 213–217.

Liu, L., Chen, C., Zhao, Y., & Zhao, E. (2015). China's carbon-emissions trading: Overview, challenges and future. *Renewable and Sustainable Energy Reviews*, *49*, 254–266.

Lo, A. Y. (2016). Challenges to the development of carbon markets in China. *Climate Policy*, *16*(1), 109–124.

Löfgren, Å, Wråke, M., Hagberg, T., & Roth, S. (2014). Why the EU ETS needs reforming: An empirical Analysis of the impact on company investments. *Climate Policy*, *14*(5), 537–558.

Martinich, J., & Crimmins, A. (2019). Climate damages and adaptation potential across diverse sectors of the United States. *Nature Climate Change*, *9*, 397–404.

Mathews, J. A., & Tan, H. (2016). Circular economy: Lessons from China. *Nature*, *531*(7595), 440–442.

Niu, S., Liu, Y., Ding, Y., & Qu, W. (2016). China's energy systems transformation and emissions peak. *Renewable and Sustainable Energy Reviews*, *58*, 782–795.

Perino, G. (2018). New EU ETS phase 4 rules temporarily puncture waterbed. *Nature Climate Change*, *8*(4), 262–264.

Raza, A., Razzaq, A., Mehmood, S. S., Zou, X., Zhang, X., Lv, Y., & Xu, J. (2019). Impact of climate change on crops adaptation and strategies to tackle its outcome: A review. *Plants*, *8*(2), 34.

Robinson, S. A., Blair, C., & McDonough, M. (2021). China's climate ambition: Revisiting its first nationally determined contribution and centering a just transition to clean energy. *Energy Policy*, *155*, 112350.

Sanin, M. E., Violante, F., & Mansanet-Bataller, M. (2015). Understanding volatility dynamics in the EU-ETS market. *Energy Policy*, *82*, 321–331.

Shan, Y., Guan, Y., Hang, Y., Zheng, H., Li, Y., Guan, D., & Hubacek, K. (2022). City-Level emission peak and drivers in China. *Science Bulletin*, *67*(18), 1910–1920.

Tang, R., Guo, W., Oudenes, M., Li, P., Wang, J., Tang, J., & Wang, H. (2018). Key challenges for the establishment of the monitoring, reporting and verification (MRV) system in China's national carbon emissions trading market. *Climate Policy*, *18*(sup1), 106–121.

Tollefson, J. (2021). COVID curbed 2020 carbon emissions-but not by much. *Nature*, *589*, 343.

Tong, D., Zhang, Q., Zheng, Y., Caldeira, K., Shearer, C., Hong, C., Qin, Y., & Davis, S. J. (2019). Committed emissions from existing energy infrastructure jeopardize 1.5°C climate target. *Nature*, *572*(7769), 373–377.

Weikmans, R., Asselt, H. V., & Roberts, J. T. (2020). Transparency requirements under the Paris agreement and their (un) likely impact on strengthening the ambition of nationally determined contributions (NDCs). *Climate Policy*, *20*(4), 511–526.

Zhang, Z. (2015). Carbon emissions trading in China: The evolution from pilots to a nationwide scheme. *Climate Policy*, *15*(sup1), S104–S126.

Zhang, D., Karplus, V. J., Cassisa, C., & Zhang, X. (2014). Emissions trading in China: Progress and prospects. *Energy Policy*, *75*, 9–16.

Zhang, L., Qin, Q., & Wei, Y. M. (2019). China's distributed energy policies: Evolution, instruments and recommendation. *Energy Policy*, *125*, 55–64.

Zhang, Y. J., & Wei, Y. M. (2010). An overview of current research on EU ETS: Evidence from its operating mechanism and economic effect. *Applied Energy*, *87*(6), 1804–1814.

Zhang, P., Yang, Y., Shi, J., Zheng, Y., Wang, L., & Li, X. (2009). Opportunities and challenges for renewable energy policy in China. *Renewable and Sustainable Energy Reviews*, *13*(2), 439–449.

Zheng, S., Yi, H., & Li, H. (2015). The impacts of provincial energy and environmental policies on air pollution control in China. *Renewable and Sustainable Energy Reviews*, *49*, 386–394.

2 Effects of Low Emission Strategies

2.1 Role of cities in combating climate change

Mitigating global warming through carbon emission reduction has received top priority in global environmental policy agenda (Fang & Heijungs, 2015). The Paris Agreement reached in December 2015 is committed to keeping warming below 2°C and even below 1.5°C above pre-industrial levels (UNFCCC, 2015). In response to the commitments on climate change mitigation, cities that contribute to three-quarters of global energy consumption and greenhouse gas (GHG) emissions are taking a critical role in addressing mitigation actions and policies (Bruestle, 1993; Hamin & Gurran, 2009; Liu et al., 2012; Xu et al., 2012). However, the effects of a series of actions and policies on climate change mitigation have not been thoroughly examined at the city scale, and low-carbon promotion at urban level requires intensive academic attention (Dong et al., 2013a, 2014b).

Urban transition provides innovative pathways for cities to realize sustainable development and address the challenges for climate change. It is well known that the urban transition consists of different stages for urban development, starting from the poverty stage, production stage, consumption stage, and finally to a pursuit of the eco-city stage. Among these challenges, how to coordinate the relationship of industry, urban, and rural area harmoniously becomes the most critical one.

Circular economy strategies can contribute to the above transition. Via system-wide innovation, these strategies aim to redefine products and services in order to enhance resource efficiency and to design waste out, while minimizing negative impacts (Dong et al., 2013a, 2014b, 2016b). It includes, but not limited to, "3R" (i.e., to reduce, reuse and recycle), urban industrial symbiosis (i.e., the exchange of resource and waste across industries or between cities and industries) as well as green supply chains serving as one critical approach to boosting urban transition (Dong et al., 2016a; Li et al., 2015). By enhancing resource efficiency (via 3R and waste utilization) and coordinating the industries and cities (via urban industrial symbiosis), circular economy strategies could strongly support sustainable urban resource management and sustainable transition. Globally, various related approaches and policies aiming at the transition are promoted, including, but not limited to: (a) *Cleaner production*, which as micro and meso level (at company and industrial park level) environmental protection initiative, effectively minimized

DOI: 10.4324/9781003433859-2

waste and emissions via process optimization (Meng et al., 2017; Zou et al., 2017); (b) *3R promotion* in Germany and Japan, with focuses on building material sound and recycling type society 1997 (Moriguchi, 2000); (c) *eco-industrial parks*, which is a practice of industrial symbiosis, verified their resource efficiency and low-carbon effects in Europe (Denmark) (Chertow, 2000; Jacobsen, 2006), South Korea (Behera et al., 2012; Park et al., 2008), Kwinana and Gladstone, Australia (Van Beers et al., 2007), and China (Bai et al., 2014; Tian et al., 2014; Zhang et al., 2010); (d) *National industrial symbiosis project* in the United Kingdom (Costa et al., 2010; Mirata, 2004), focusing on regional waste recycling and exchanges; and (e) *Eco-town and "zero emission"* practice in Japan (Hashimoto et al., 2010; Van Berkel, 2010), in which urban industrial symbiosis and recycling industries promotion was core. Among all these promotions, urban industrial symbiosis was one key part as it optimized the urban and industrial metabolism, and particularly, its promotion significantly reduced the carbon emissions at the industrial cluster (Hashimoto et al., 2010; Van Berkel, 2010) and city level (Geng et al., 2010; Jacobsen, 2006). As such, accounting the carbon emissions and discussing the related environmental responsibilities had become the hot topic, and emerging studies provided enlightening information, e.g., the boundary setting, carbon emissions accounting for entities presented in Kim et al. (2018)

With the rapid and intertwined urbanization and industrialization, China makes it possible to trace the urban transition and the urban GHGs mitigation via circular economy promotion. By 2010, the urban population in China had increased to more than 660 million, with an urbanization rate of about 50%. These numbers are expected further up to 850 million and 60% in the coming years, respectively (Chen et al., 2013; NBS, 2011; UN, 2012). In response to the growing challenges of resource depletion and climate change, the Chinese government has promoted a series of policies to support sustainable urban development with more concerns on environmental protection and eco-civilization construction. The important policy initiatives includes, but are not limited to: the "Circular Economy" strategies at the city level, the "Eco-City" project (Liu et al., 2014a; Nan Zhou et al., 2012; Wu, 2012), and the "Low-Carbon City" project (Su et al., 2012; Yang & Li, 2013). In all the projects, the circular economy remains the key in view of its considerable effectiveness on the enhancement of resource efficiency and reduction in waste generation.

As the world's largest GHG emitter, China has made a remarkable contribution to the negotiation of the Paris Agreement by setting its ambitious carbon emissions reduction target toward 2030 (Zhao et al., 2017) and speeding up the development of circular economy (Jiao & Boons, 2017). China is also one of the countries that make "circular economy" as national strategy (Dong et al., 2013b). A number of related projects have been promoted nationally and regionally, such as the "National Eco-Industrial Park" project by the Ministry of Environment Protection (MEP) and "Circular Economy Pilots Projects" by the National Development and Reform Commission (NDRC) (Dong et al., 2014b, 2016a). However, the evidence of low-carbon benefit of a circular economy is less reported in China, particularly at the urban level. While the carbon footprint (CFP) serves as a mature tool for responding to climate change concerns, limited studies have made use of CFP as a proxy for the performance of urban circular economy promotion (Dong et al., 2014a, 2016b). Besides, the quantitative evidence on how the circular economy practices lead to reductions in carbon emissions is less reported, let alone the evaluation at the city level.

CFP, which is a fundamental member of the footprint family (Fang et al., 2016), offers the opportunity scientifically to account for the direct and indirect carbon emissions associated with human activities, normally subject to a consumption-based accounting. It has attracted tremendous interests from academia, the public, organizations and governments as an approach for informing efforts toward low-carbon societies (Kanemoto et al., 2016; Wiedmann, 2009). The CFP of a city can be defined as the amount of CO_2-equivalent emissions caused directly and indirectly by human activities within the urban system boundaries. To provide an overall picture of the state-of-the-art in CFP studies, we conducted a literature search through Science Citation Index Expanded during 2007–2016, and find there are 2545 publications with keywords "carbon footprint," "carbon footprints," or "carbon footprinting." Those words that have the same or similar meaning have been grouped into one word. For example, life cycle assessment, life cycle analysis, LCA, life-cycle assessment, and life-cycle-assessment are all classified as "life cycle assessment." Apparently the most prominent words highlighted are "carbon footprint" in itself. This is followed by the words of "life cycle assessment" and "greenhouse gas (GHG) emissions." It shows that "sustainability," "climate change," "CO_2 emissions," "energy efficiency," and "industrial ecology" are also booming with high frequency of emergence for no less than 60 times.

Broadly speaking, there are three major methods to accounting for CFP, including accounting based on Intergovernmental Panel on Climate Change (IPCC), recommended emission inventory and factors (hereafter IPCC method), life cycle assessment (LCA), and input–output analysis (IOA). The IPCC method is the mostly applied one to account for the direct emissions caused by energy and land use, primarily appropriate for use in macro entities (e.g., nations). When being applied to micro-entities such as products and enterprises, the accuracy of the IPCC method may not be satisfactory. While LCA and IOA both can be used to calculate direct and indirect emissions, the former concentrates on micro entities such as products (Baldini et al., 2017), consumers (Jukka et al., 2013), and enterprises (Liu et al., 2017). Although the result figured out by the LCA method is more detailed, it still faces some limitations such as uncertainty, data unavailability, etc. By contrast, IOA is mainly used to measure the emissions of macro entities such as nations (Barrett et al., 2013), regions (Feng et al., 2013), and industries (Tidy et al., 2016), which could overcome the shortcoming of truncation error while needing input–output tables (IOTs) in which extensive aggregation of sectors may lead to inaccuracy. In keeping with continuous development and refinement of IOA and basic datasets (Minx et al., 2009), nowadays more researchers are attempting to employ IOA to analyze meso entities and, in some cases, even micro-entities in combination with LCA (e.g., hybrid LCA). Nevertheless, it has been observed that the results of hybrid LCA are not necessarily more accurate than those of traditional process-based LCA unless the input–output (IO)-based LCA and process-based LCA have the same level of detail (Yang et al., 2017). Besides, given the limited coverage of hybrid LCA accounts (typically including food, water, fuels, and cement), the estimation may not be adequately informative since many other materials are produced out of the cities but consumed within them (Ramaswami et al., 2008).

Despite the increasing studies on the CFP accounting of nations and regions by means of IOA, studies on the CFP accounting of cities remain largely unexplored, with a few exceptions such as extended territorial carbon emissions in UK's cities (Jan et al., 2013), Melbourne's carbon map (Wiedmann et al., 2016) and consumption-based carbon emissions of China's 13 cities (Mi et al., 2016), probably because of the lack of IOTs of cities and the mismatching between financial and physical flows (Ramaswami et al., 2011). The first application of IOA at the city level was to Denver City, for which the IPCC method has been found incapable of addressing indirect carbon emissions embodied in traded goods. Nowadays, there have been a growing number of cities releasing their urban GHG emission inventories, but most of these overlook indirect emissions embodied in imports and exports (Ramaswami et al., 2008). Only few urban inventories include some items related to indirect carbon emissions, such as those of Aspen and Seattle (Hillman & Ramaswami, 2010).

With this circumstance, this paper aims to investigate the CFP from 2002 to 2012 in Guiyang City, China. Guiyang is highlighted as an ideal laboratory to test the effects of circular economy promotion engaged in a number of related national pilots project, including: a. China's Circular Economy City pilot (since 2005) by NDRC; b. Low-Carbon City pilot (since 2011) by NDRC; c. Ecological Civilization City pilot (since 2014) by MEP. It hence offers a good case to verify our developed approach and to seek for the follow-up implications. To support this analysis, a hybrid model integrating IO approach and process-based inventory analysis is developed to investigate the urban CFP and the impacts of various circular economy scenarios (the year of 2002-business as usual (BAU) scenario, 2007-initially applied the circular economy strategy since the pilot in 2005 and 2012-circular economy promotion enhanced after years' practice). However, given the lack of IOTs of Guiyang City, we are about to combine the IOTs of Guizhou Province with fiscal data of Guiyang City with the same aggregation of the details, which could minimize the deviation of calculation and depict the direct and indirect carbon emissions explicitly in comparison to the IPCC method. Furthermore, the IOA method will allow us to disclose the nexus between direct and indirect sectoral carbon emissions both from production and consumption perspectives, thus pointing to the hot-spot industries of a region in a walk. Particularly, we identify scenarios related to proposed urban industrial symbiosis. Analytical results are expected to provide critical insights into the role of circular economy in speeding up urban transition toward a low-carbon society. Policy recommendations aimed to tackle the barriers to regional decarbonization are also proposed in support of wise decision-making. We believe that this book will be informative for policymakers of urban planning by shedding light on innovative eco-industrial development and urban transition in China.

The remainder of this paper is organized as follows: Section 2.2 describes the methodology and data; Section 2.3 overviews the case area; Section 2.4 presents the analytical results and discussion; and finally, Section 2.5 draws the conclusions and implications.

2.2 Methods and data

2.2.1 Hybrid model

As stated, we take the IOA to measure the direct and indirect emissions associated with anthropogenic activities. IOA is a typical top-down approach that reflects the relationship of initial, intermediate, and total inputs, as well as the outputs of the intermediate, final, and total. This approach was first proposed by Leontief in 1936 and was proven successful in predicting the USA's demand for steel in 1950 (Leontief, 1936). Afterward, this method was not only for economic issues but gradually extended to measure the embodied resource, energy, and emissions by researchers from a vast range of fields.

When choosing IOA for the city-wide CFP accounting, we must adjust the common pattern of this method compared to macro-level cases mostly due to the lack of sound IO data. Three components can be identified in urban CFP accounting. Simply put, the environmental coefficient matrix quantifies the carbon emissions per unit, the production structure matrix represented by IOTs reveals the productive relationship between all sectors provided in the IOTs, and the final demand matrix denotes the final consumption including imports while excluding the exports in the market (Figure 2.1).

The production structure matrix that is derived from IOTs, is a key element of IOA, expressing the complex economic relations of supply and demand between various sectors. To measure the carbon emissions from both production and consumption perspectives, we first account for the total output as follows:

$$AX + Y = X \tag{2.1}$$

$$A = \begin{bmatrix} a_{11} & a_{12} & \cdots & a_{1n} \\ a_{21} & a_{22} & \cdots & a_{2n} \\ \vdots & \vdots & \ddots & \vdots \\ a_{n1} & a_{n2} & \cdots & a_{nn} \end{bmatrix} \tag{2.2}$$

where A is the production coefficient matrix, which is meant to show the direct input of sector i into sector j; X is the total output vector; AX represents the intermediate

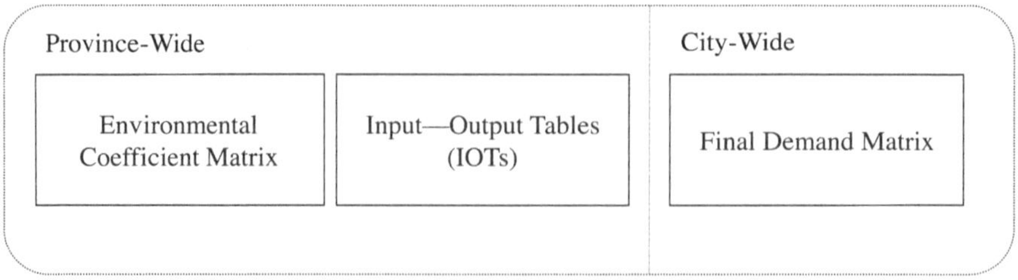

Figure 2.1 Schematic of urban carbon footprint accounting based on IOA

output; Y is the final demand vector; x_{ij} is the input x of sector i into sector j; and i and j represent for different sectors.

Afterward, Equation (2.1) can be transformed as follows:

$$(I - A) \cdot X = Y \tag{2.3}$$

$$X = (I - A)^{-1} \cdot Y \tag{2.4}$$

$$Y = \begin{bmatrix} y_1 & y_2 & \cdots & y_n \end{bmatrix}^T \tag{2.5}$$

$$X' = (I - A)^{-1} \cdot \widehat{Y} \tag{2.6}$$

where $(I - A)^{-1}$ is the Leontief inverse matrix, which is to allocate the overall demands among different sectors in production; $\widehat{Y}$ is the diagonal matrix of Y representing the final demand of the market; y_i is the final demand of sector i; and X' differs from vector X in that X' represents the detailed demands of each sector. By doing so, we have measured the intermediate and total demands of each sector in the X' matrix, and disclosed the economic relationship among all the sectors involved. Furthermore, we translate the economic relationship to the relationship of carbon emissions with the environmental coefficient matrix as follows:

$$C = \begin{bmatrix} c_1 & c_2 & \cdots & c_n \end{bmatrix} \tag{2.7}$$

$$E = \begin{bmatrix} e_{11} & e_{12} & \cdots & e_{1n} \\ e_{21} & e_{22} & \cdots & e_{2n} \\ \vdots & \vdots & \ddots & \vdots \\ e_{n1} & e_{n2} & \cdots & e_{nn} \end{bmatrix} = \widehat{c} \cdot X' = \widehat{c} \cdot (I - A)^{-1} \cdot \widehat{Y} \tag{2.8}$$

where C is a carbon intensity vector representing the carbon emissions per unit; $\widehat{c}$ is the diagonal matrix of C; and E is the carbon emission matrix of all sectors with e_{ij} representing the carbon emissions by sector i due to the demand of sector j. Due to data availability, we consider CO_2 emitted from the entity merely. To calculate the carbon emissions from both production and consumption perspectives, two equations are used to measure the carbon emissions of different sectors in a comparative sense

$$\mathrm{CE}_{p,i} = \sum_{j=1}^{n} e_{ij} (i = 1,2,3,\cdots) \tag{2.9}$$

$$CE_{c,j} = \sum_{i=1}^{n} e_{ij} (j = 1,2,3,\cdots) \tag{2.10}$$

where $CE_{p,i}$ is carbon emissions of sector i based on a production perspective, which is the sum of each row of matrix E; and $CE_{c,j}$ is carbon emissions of sector j embodied in consumption, which is the sum of each column of matrix E.

Furthermore, we define Indirect/Direct Ratio Index (IDRI) as a measure of dividing the indirect carbon emissions by the direct to quantify the distinction between direct and indirect carbon emissions in different sectors both from production and consumption perspectives, thus distinguishing carbon emissions from the internal and external, and clarifying the responsibility of individual entities to some extent respectively from a production or consumption perspective (Equations 2.11 and 2.12). Likewise, to explore the relationship of carbon emissions among various sectors both from production and consumption perspectives, a new indicator named Production/Consumption Ratio index (PCRI) is developed as a measure of dividing carbon emissions from a production perspective by that from a consumption perspective (Equation 2.13). The PCRI focuses on the carbon emissions of the supply chain by contrasting the upstream and downstream, thus could be informative for policymakers to optimize the composition of the supply chain

$$\mathrm{IDRI[P]}_i = \frac{e_{ii}}{CE_{p,i}} (i = 1,2,3,\cdots) \tag{2.11}$$

$$\mathrm{IDRI[C]}_i = \frac{e_{ii}}{CE_{c,i}} (i = 1,2,3,\cdots) \tag{2.12}$$

$$\mathrm{PCRI}_i = CE_{p,i} / CE_{c,i} (i = 1,2,3,\cdots) \tag{2.13}$$

where P and C represent the carbon emissions from a production and consumption perspective, respectively.

2.2.2 *Scenarios design*

To test the approach feasibility and highlight policy implications, scenarios of urban industrial symbiosis are designed for case analysis

2.2.2.1 Supply and demand match for symbiotic network design

Supply and demand match is conducted as the basis for setting urban industrial symbiosis scenarios. Potential of material/waste exchange between enterprises and or sectors will be analyzed with technology feasibility analysis (to see whether two parts can be linked with material/energy/waste flows from an engineering perspective); if not, the symbiotic design will be denied. If yes, the amount of exchanged flows will depend on the analysis on both the resource capacity and the feasibility of material/energy flows synergies (technological feasibility). An analytical scheme for supply–demand match is illustrated in Figure 2.2. When the material (and or wastes/energy) flows (W_{ij}) of upstream enterprise i cannot satisfy

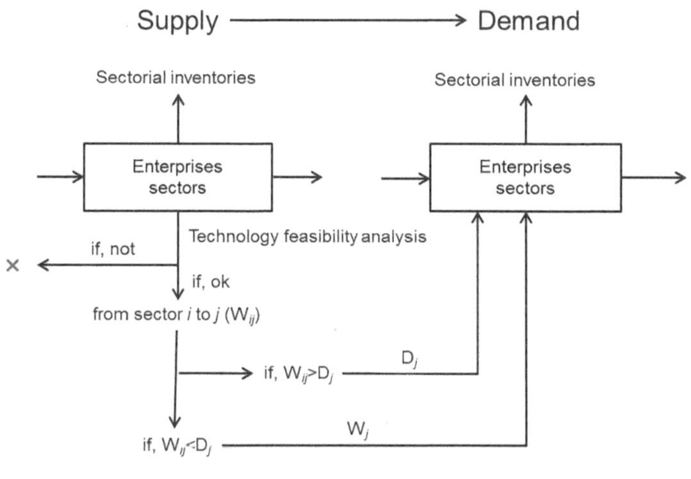

Figure 2.2 Scheme of supply and demand match

Note: The arrows refer to material/energy/waste flow.

the feasible demand (D_j) in downstream enterprises j, and the exchange flow is set into W_j. In another condition, when the flows (W_{ij}) of upstream enterprise i are over the feasible demand (D_j) in downstream enterprises, the flows of synergy are set as D_j. The whole symbiotic network is designed in this way. The related resource and technological capacity are referred according to survey and expert consulting to engineers in related companies via our project (described in data acquisition).

2.2.2.2 *Scenarios setting*

According to the above scheme, scenarios are set based on local condition. China's IOTs (the basis for the established CFP approach) has been published every five years (e.g.,. in the year of 1997, 2002, 2007, and 2012, which is the most updating one); hence, our analytical scenarios basically follow the 5-year time span. In addition, from 2000 to 2014, several milestones happened achieved in the environmental perspective protection in our case city, Guiyang (more about the case will be described in the next section), including that (a) it became China's Circular Economy City pilot in 2005; (b) it became the national Low-Carbon City pilot in 2011; and (c) it became the Ecological Civilization City pilot in 2014. Such promotions were reflected in the change of resource efficiency with economic growth (GDP per capita), e.g., domestic material consumption (DMC)/capita. Based on this, our scenarios are designed as follows:

a BAU scenario: in the year of 2002. According to Figure 2.6, its GDP/capita and DMC/capita were increasing quickly;
b Urban development after becoming the circular economy pilot: the year of 2007, two years after becoming the circular economy pilot;

c Urban development after years of circular economy promotion: the year of 2012, seven years after becoming the circular economy pilot and 1 year after becoming the Low-Carbon City pilot;
d Urban industrial symbiosis (UIS) scenario in 2012 (2012-UIS): to enhance the circular economy promotion, Guiyang has implements implemented the UIS planning in 2011, and t. This scenario considers several selected UIS options.

As to the 2012-UIS scenario, the UIS is designed based on local condition, which is illustrated as Figure 2.7. The key synergies designed include:

a Industrial solid waste exchange: slag from iron/steel industry, aluminum industry, and phosphorous chemical industry, slag as well as coal gangue from power plant and coal industry to cement industry;
b Traditional recycling: like waste steel recycling;
c Municipal solid waste that is able to be an energy resource produces energy sources to substitute fossil fuels (like scrap tires and waste plastics) be recycled into the furnaces of iron and steel sector;
d Energy symbiosis: waste heat utilization between industrial sectors (from iron/steel to power generation; from aluminum processing to power generation; from power plant to nearby industrial plants), and industrial waste heat provision to urban areas (e.g., commercial and residential buildings) and agriculture sector (e.g., steam provision to greenhouse).

As complementary information for the scenarios, the general environmental benefits (direct analysis) are summarized in Table 2.1. More detailed information about the UIS can be found in our previous studies. Based on the analysis of the environmental benefits, CFP can be further evaluated.

2.2.3 Data

In view of the lack of IOT for Guiyang, accounting for Guiyang's CFP would require a harmonized combination of province-wide and city-wide datasets, as stated in Figure 2.4. To that end, an updated IOA method will be employed to combine the IOTs of Guizhou Province with fiscal data from Guiyang City, both of which are aggregated to the same level of detail. To contrast the carbon emissions both from production and consumption perspectives, we are about to combine the IOTs of 2002, 2007, and 2012 for Guizhou Province with the data for the consumption of various types of energy of this province and data for final demands of Guiyang each year, which offers the opportunity to explore the effects of China's strategy perfectly through the reflection of production technology changing in 2002 (the baseline), 2007 (after becoming the circular economy pilot), and 2012 (with the implementation of UIS). The IOTs for Guiyang Province in 2002, 2007, and 2012 are derived from the Guizhou Bureau of Statistics containing 42 economic sectors, while the final demands of Guiyang are from the Guiyang Statistic Yearbook (GSY) of 2002, 2007, and 2012. Since China does not officially release the carbon

Table 2.1 Resource savings and waste reductions through industrial and urban symbiosis in Guiyang

	Symbiosis	*General environmental benefits*
Industrial solid waste utilization	Substitute cement material with steel slag (120 kt/y)	Save material of clinker 120 kt/y and reduce the process CO_2 emissions by 66 kt/y. Reduce slag by 120 kt/y.
	Substitute cement material with phosphorus slag (500 kt/y)	Save material of clinker 500 kt/y and reduce the process CO_2 emissions by 275 kt/y. Reduce slag by 500 kt/y.
	Substitute cement material with aluminum industrial slag (400 kt/y)	Save material of clinker 400 kt/y and reduce the process CO_2 emissions by 220 kt/y. Reduce slag by 400 kt/y.
	Coal gangue utilized by power generation (100 kt/y)	Save fossil fuel by 30 ktce/y. Reduce CO_2 emissions by 78 kt/y. Reduce solid waste by 100 kt/y.
	Coal flying ash recycling (200 kt/y)	Reduce solid waste. Save raw material of cement and reduce the process CO_2 emissions by 200 kt/y. Reduce ash by 200 kt/y.
Tradition recycling	Waste steel recycling (100 kt/y)	Save raw material 2.5 t and energy 12.25 GJ/t steel. Reduce CO_2 emissions by 108.8 kt/y.
Municipal waste to industries	Waste plastics recycling (10 kt/y)	1 t waste plastic could substitute 1.2 t coke. Reduce CO_2 emissions by 31.2 kt/y.
	Scrap tires recycling (10 kt/y)	1 t scrap tire could produce 0.65 tons rubber powder and consume 0.04 tce electricity.
Energy symbiosis	Heat exchange (300 t/y)	Save fossil fuel by 18,864 tce and reduce CO_2 emissions by 49,046 t/y.

Note: 1 tce = 29.27×10^3 GJ; kt = 10^3 ton. 1 t cement production needs 1.4 t ore mining (based on the survey on local cement company).

intensity of sectors and the details on provincial data are insufficient, especially in the early period of this research, the IPCC method will be employed to yield these parameters, together with the Energy Statistic Yearbook (ESY) of Guizhou Province published for the same year.

As extensive sectors aggregation may lead to significant inaccuracy of the accounting, data availability remains a great barrier for disaggregation of energy consumption data, especially at the early years. To harmonize the sectorial categorization in the IOTs and ESY and allow them to have different levels of details with Guiyang GSY, we aggregate the 42 sectors provided by IOTs into five aggregated sectors, namely Agriculture and Forestry, Industry, Construction, Transportation, Storage, and Postal Service, and Wholesale and Retail. Note that this kind of aggregation is common and effective at reducing the deviation as far as possible (Lin et al., 2013), thus yielding the best datasets to our knowledge (Su et al., 2010; Zhang et al., 2015). Furthermore, to explore the efficiency of industrial transformation mostly from the second to the third and disclose the

Table 2.2 Direct consumption coefficient matrix in 2002, 2007, and 2012

	2002					*2007*					*2012*				
	A	*B*	*C*	*D*	*E*	*A*	*B*	*C*	*D*	*E*	*A*	*B*	*C*	*D*	*E*
A	0.17	0.09	0.00	0.00	0.03	0.15	0.04	0.00	0.02	0.01	0.14	0.04	0.00	0.02	0.01
B	0.13	0.42	0.57	0.31	0.15	0.15	0.49	0.59	0.16	0.15	0.17	0.49	0.56	0.23	0.14
C	0.00	0.00	0.00	0.00	0.01	0.00	0.00	0.00	0.01	0.01	0.00	0.00	0.01	0.02	0.01
D	0.02	0.06	0.05	0.02	0.03	0.02	0.06	0.05	0.04	0.04	0.03	0.06	0.05	0.10	0.05
E	0.03	0.09	0.14	0.10	0.22	0.04	0.10	0.12	0.31	0.18	0.05	0.09	0.13	0.17	0.20

Note: A represents Agriculture and Forestry; B represents Industry; C represents Construction; D represents Transportation, Storage, and Postal Service; and E represents Wholesale and Retail.

inner relationship of carbon emissions among industries in production and consumption, we also aggregate those five sectors into three domains (Agriculture, Industry, and Service), which could offer a policy-oriented insight to the reveal the overall carbon map among industrial structure transformation in Guiyang or even in China. Accordingly, the final demands derived from GSY follow the same schemes for aggregation; that is, the aggregation of three sectors consists of Agriculture, Industry, and Service, and the aggregation of five sectors consists of Agriculture and Forestry, Industry, Construction, Transportation, Storage, and Postal Service, and Wholesale and Retails.

Besides, given the composition of national GHG inventory of China, we will replace the data for energy consumption with those for the consumption and carbon intensity of standard coals in calculating the carbon intensity of each sector. Therefore, the direct consumption coefficient matrix in 2002, 2007, and 2012 as well as the GDP of five sectors are provided in Tables 2.2 and 2.3.

2.3 Introduction to case city

Guiyang City, the capital of Guizhou Province, is one of China's national pilots of the Circular Economy City and the Low-Carbon City. More recently, it has launched a new strategy toward an Eco-Civilized City in support of China's circular economy actions. The situation is, however, that Guiyang's economy has still

Table 2.3 GDP of each sector in 3 years

	GDP (100 million yuan)		
	2002	*2007*	*2012*
A	41.01	71.55	62.55
B	104.99	248.90	454.90
C	34.68	71.57	121.94
D	22.27	35.64	187.72
E	255.39	642.65	733.68

Note: A represents Agriculture and Forestry; B represents Industry; C represents Construction; D represents Transportation, Storage, and Postal Service; and E represents Wholesale and Retail.

been depending heavily on the inputs of fossil energy, particularly coals, over past decades (Dong et al., 2016a; Li et al., 2015). Even though Guiyang's economic output of service in terms of GDP has exceeded that of industry recently, equipment manufacturing, chemical industry, aluminum processing, modern medicine, specialty food, and tobacco products are still dominating sectors that require a great amount of energy input (Liang & Jin, 2011). All this has posed a serious challenge to the low-carbon urban transition. For this reason, it is necessary to evaluate the performance of circular economy practices on carbon emission reductions by investigating the dynamics of Guiyang's CFP (Liu et al., 2014b).

Catching up with the national "environmental legislation" campaign since 2000, Guiyang has implemented atmospheric quality improvement project aimed at making use of cleaner energy as well as formulating strict environmental standards for enterprises and organizations, altogether with several other projects toward water governance and green belt rebuilt. Afterward, Guiyang was selected as a Circular Economy Pilot City in 2005 and as a Low-Carbon Pilot City in 2011. In line with the implementation of these strategies raised in Guiyang, some sustainable development goals have been achieved gradually, such as the fight against acid rain. However, it remains questionable whether the effect of the circular economy strategy is able to continue benefiting the population after 2012 and in the future.

In terms of resource exploration and consumption, the resource consumption of Guiyang (we present the data of DMC in the categories of metal ores, nonmetal ores, fossil fuels, and biomass) had increased from 15.3 million tons to 51.7 million tons, nearly 3.5 times changed. From an efficiency perspective, the resource productivity (economic output per unit of resource consumption), which was a key resource efficiency indicator for measuring the circular economy by NDRC (Dong et al., 2016a), had significantly increased, from 1700 to 4800 CNY/t. However, compared with the resource productivity of developed countries (here we compared with Japan and South Korea as Asian members), there still has to be ample room for improvement. In terms of wastes generations, industrial waste as well as mining waste, like the tailings were a big issue for Guiyang. Although the absolute amount kept increasing in general trend, the comprehensive utilization rate for industrial solid waste had been improved as a result of continuous circular economy practices. The next step for Guiyang would be to deepen the circular economy implementation, particularly to enhance the UIS practice to further improve the bulky waste exchange among industrial sectors and urban areas.

The following analysis will trace the benefits of urban environment improvements in Guiyang and the impacts of UIS on urban CFP reductions based on the hybrid approach we constructed in the previous section.

2.4 Results and discussion

2.4.1 Scenarios analysis of urban carbon footprint

Based on the scenarios setting, we systematically investigate and compare the urban CFP in 2002, 2007, and 2012 to trace the CFP mitigation along with urban development.

Table 2.4 Comparison of carbon intensity in 3 years

	Carbon intensity (tCO_2/10 000 yuan)		
	2002	2007	2012
A	1.50	1.13	0.24
B	2.89	4.35	1.20
C	0.59	0.25	0.05
D	2.09	2.44	0.63
E	0.65	0.76	0.74

Note: Data for fossil energy consumption and GDP derive from Guizhou ESY and GSY, and data for emission inventories and factors derive from IPCC.

2.4.1.1 *Carbon intensity of sectors*

The carbon intensity of Industry, Transportation, Storage, and Postal Service, and Wholesale and Retail increased between 2002 and 2007 (Table 2.4), among which the carbon intensity of Industry surged as a result of rapid industrialization and urbanization. Afterward, the carbon intensity of all sectors went down notably during 2007–2012, in which circular economy promotions were fully implemented, with the exception of Wholesale and Retail, whose carbon intensity fluctuated slightly over the same period of time. Supporting data is presented in Appendix A, Tables A.1 and A.2.

2.4.1.2 *Carbon footprint of sectors*

With respect to the three-sector classification, Industry always occupied the largest proportion of the total CFP while Agriculture took the least (Figure 2.3). The proportion of Industry in production-based carbon emissions increased from 11.1 million tCO_2 in 2002 to 35.3 million tCO_2 in 2007 and then decreased to 18.7 million tCO_2 in 2012. Trends in Agriculture and Service were on the contrary, ranging from "1.53-1.35-0.39" million tCO_2 and from "3.9-4.9-9.4" million tCO_2, respectively. Furthermore, the proportion of Industry in production-based carbon emissions was always much higher than that in consumption-based carbon emissions in the three years, while Service was the other way around. Compared with these two sectors, changes to Agriculture were tiny since its proportion in production-based and consumption-based carbon emissions was quite similar.

When it comes to the five-sector classification (Figure 2.4), Industry took the largest proportion both in the production-based and consumption-based carbon emissions. During 2002–2012, trends in consumption-based carbon emissions of all the five sectors were similar, but the proportions varied widely, for Agriculture ranging from "1.53-1.34-0.39" million tCO_2, for Industry ranging from "11.1-35.4-18.6" million tCO_2, for Construction ranging from "0.02-0.1-0.1" million tCO_2, for Transportation, Storage, and Postal Service ranging from "1.3-2.5-2.3" million tCO_2, and for Wholesale ranging from "2.6-2.5-7.0" million tCO_2. Agriculture and Forestry is the only sector whose proportion in both production-based and consumption-based carbon emissions kept falling. The proportion of Transportation,

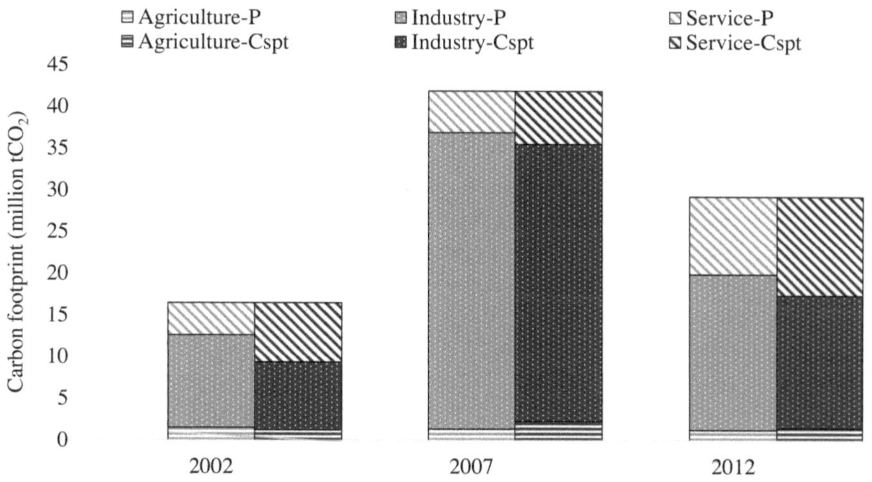

Figure 2.3 Comparison of carbon footprint between production and consumption perspectives aggregated to three sectors

Note: P represents carbon emissions from a production perspective; and Cspt represents carbon emissions from a consumption perspective (the same as below).

Storage, and Postal Service and Wholesale was always close to zero when it comes to the production-based carbon emissions, which was very different from the case of the consumption-based carbon emissions. We also find that the proportion of Agriculture and Transportation, Storage, and Postal Service failed to differ not so much between production and consumption perspectives.

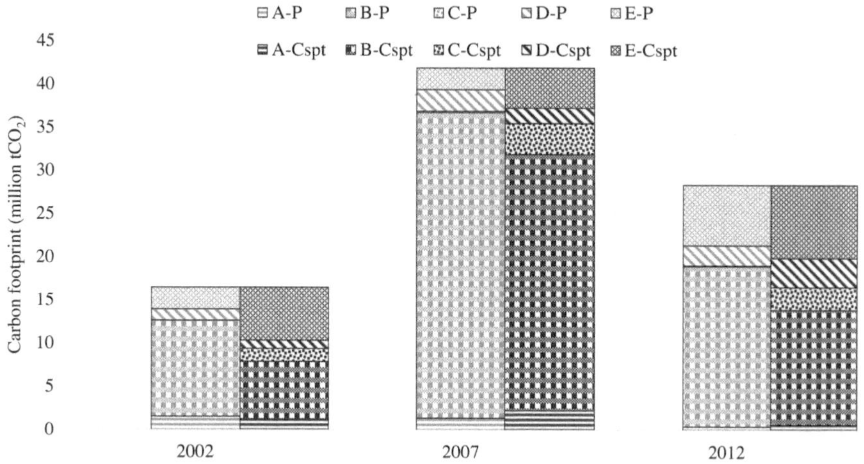

Figure 2.4 Comparison of carbon footprint between production and consumption perspectives aggregated to five sectors

2.4.2 Comparison between production and consumption perspectives

The total CFP of Guiyang City changed from 16.5 million tCO_2 in 2002, to 41.9 million tCO_2 in 2007, and to 28.4 million tCO_2 in 2012. Thus the amount of CFP in 2007 was 60.6% higher than that of 2002 and 43.3% higher than that of 2012. With the five-sector classification, we find the gap between the direct and indirect carbon emissions from both production and consumption perspectives to be considerably large (Figure 2.5(a)–(c)). The results of the two indicators (IDRI and PCRI) are presented in Table 2.5, showing that the direct carbon emissions from Agriculture and Forestry, Industry, and Construction all experienced an increase in 2002–2007 and a decrease in 2007–2012 and the same as the indirect carbon emissions from Industry, Transportation, Storage, and Postal Service from a production perspective. In addition, the direct carbon emissions from Transportation, Storage, and Postal Service based on production rose all the time, and the indirect ones of Construction and Wholesale based on consumption rose as well. As for the indirect carbon emissions from a consumption perspective, those of Agriculture and Forestry, Industry, and Construction went up and down while those of Transportation, Storage, and Postal Service always increased during 2002–2012.

In addition, the dynamics of CFP per GDP and per person were tracked and presented in Figure 2.6. Both of these two indicators increased from 2002 to 2007 and decreased afterward.

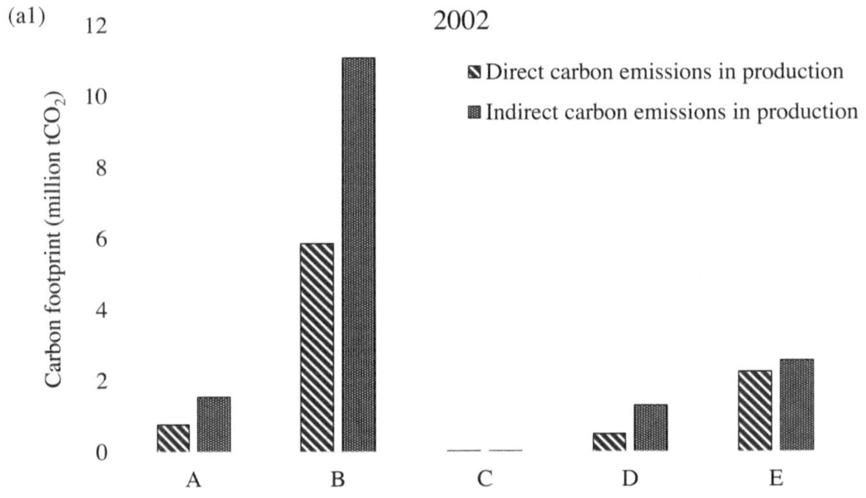

Figure 2.5 Comparison of direct and indirect carbon emissions from production and consumption perspectives. (a1–a3) Comparison of direct and indirect carbon emissions from a production perspective with five-sector classification. (b1–b3) Comparison of direct and indirect carbon emissions from a consumption perspective with five-sector classification. (c1–c3) Comparison of carbon emissions from production and consumption perspectives with five-sector classification (*Continued*)

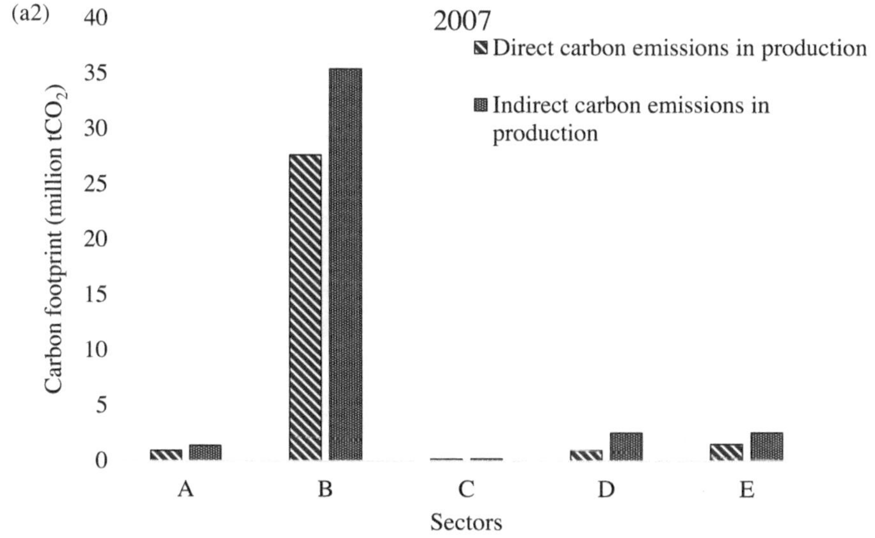

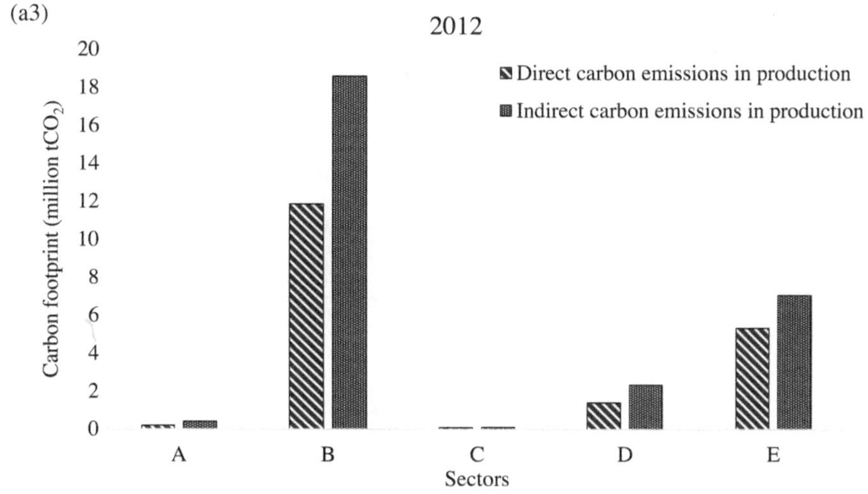

Figure 2.5 (Continued)

2.4.3 Extra carbon footprint reduction benefits from urban industrial symbiosis

UIS has been proved to be able to offer additional environmental benefits via waste exchange, energy recovery, and other resource efficiency options. This section will further investigate their impacts on the mitigation of urban CFP.

First of all, a general analysis on the direct effects (resource-saving, waste mitigation, and pollutants reduction via direct material flow analysis) of the symbiosis

(b1)

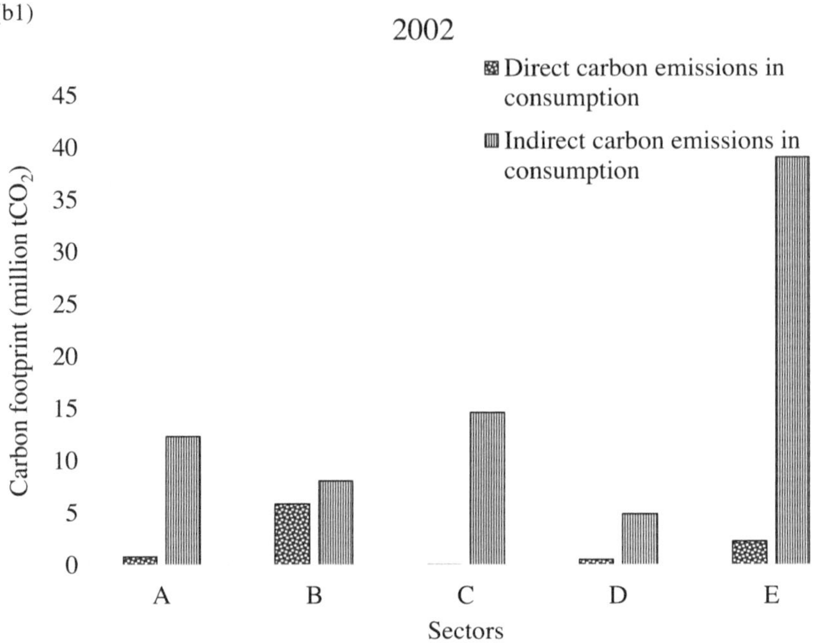

(b2)

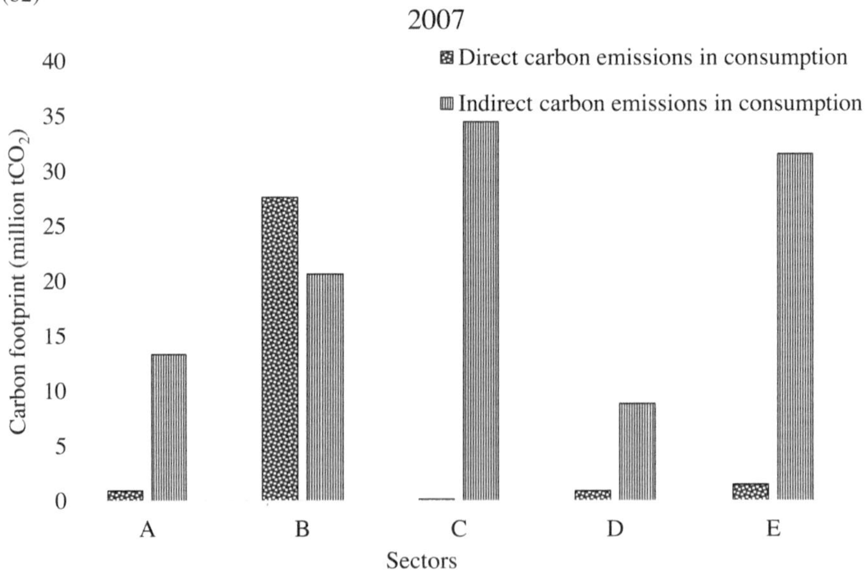

Figure 2.5 (Continued)

(b3)

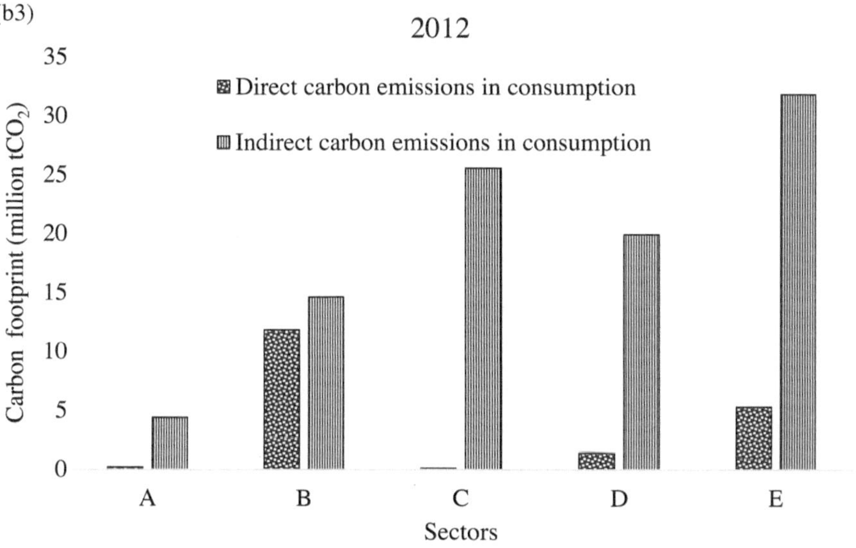

(c1)

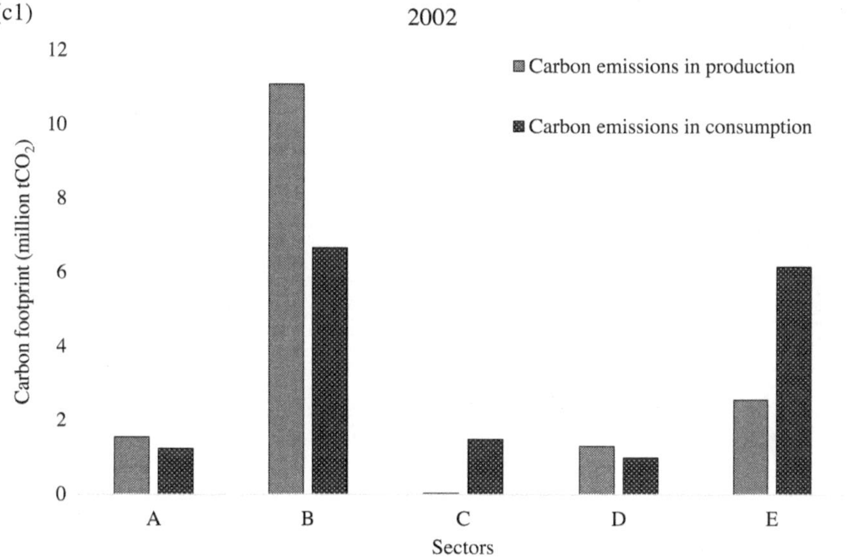

Figure 2.5 (Continued)

(c2)

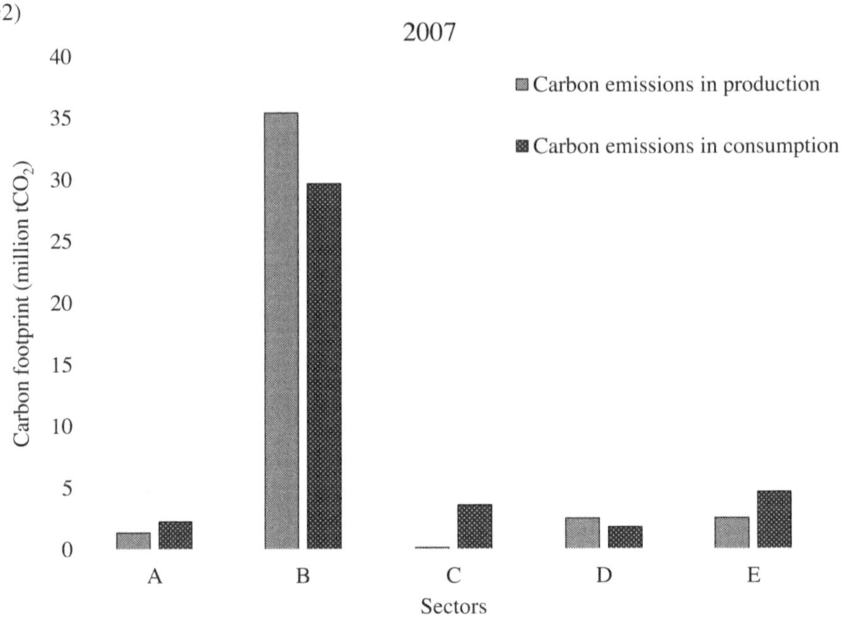

(c3)

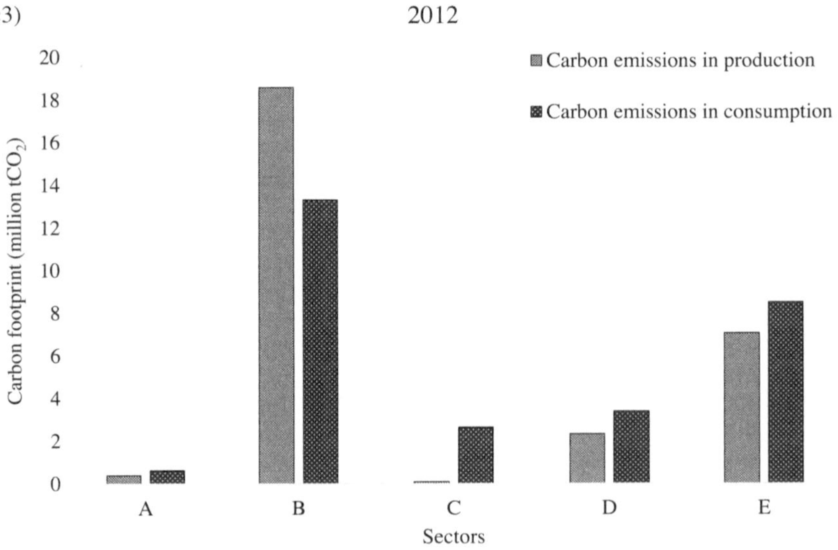

Figure 2.5 (Continued)

Table 2.5 IDRI and PCRI of sectors in 3 years

	2002			*2007*			*2012*		
	IDRI[P]	*IDRI[C]*	*PCRI*	*IDRI[P]*	*IDRI[C]*	*PCRI*	*IDRI[P]*	*IDRI[C]*	*PCRI*
A	0.99	2.02	0.85	1.64	0.68	0.40	1.25	0.60	0.63
B	1.12	3.54	1.76	7.26	13.38	8.09	1.66	1.19	1.40
C	8.20	20.61	6.59	0.01	0.04	0.03	0.02	0.04	0.03
D	0.63	0.56	1.48	1.03	1.01	0.69	1.32	1.40	0.69
E	7.14	1.41	3.03	0.57	0.47	1.66	0.42	0.54	0.83

Note: IDRI, Indirect/Direct Ratio Index; PCRI, Production/Consumption Ratio Index.

is done, and the impacts on the key industrial sectors are investigated. According to the case introduction in the previous section, the main industrial sectors include "Aluminum industry," "Phosphorus chemical industry," "Iron/steel industry," "Cement sector," "Coal chemical industry" as well as "Power generation industry." They are also the key sectors highlighted in Guiyang's local document of "12th five-year plan of industrial integration" (Dong et al., 2016a).

According to the data presented in Table 2.1 and the coefficients (see Appendix A, Tables A.3 and A.4) of material consumption and wastes generation related to the above sectors, effects of resource-saving, waste reduction through the symbiosis design were presented and allocated to certain sector in Table 2.6. It was highlighted that, with urban industrial scenarios, significant resource saving, and wastes mitigation effect were achieved. In summary, iron ore (iron/steel sector) and limestone mining (cement sector) decreased by 1120.00 and 3036.00 kt/y. Benefiting from this, tailings (which were a significant issue for industrial cities like

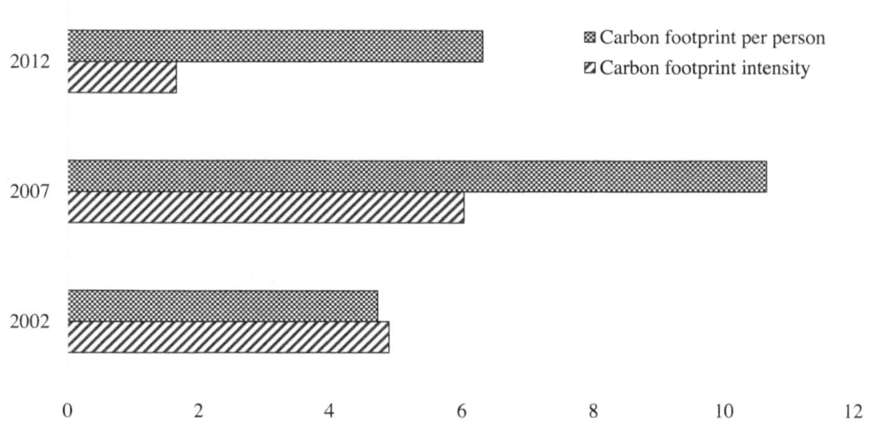

Figure 2.6 Comparison of carbon footprint intensity and carbon footprint per person

Table 2.6 Effects of resource-saving/wastes mitigation in key sectors in urban industrial symbiosis scenario

Related sectors	*Related resources/wastes*	*Effects*
Iron/steel	• Input: Iron ore (kt/y)	−1120.00
	Fossil fuel (ktce/y)	−72.37
	• Output: Slag (kt/y)	−120.00
	Tailing (kt/y)	−1950.00
Cement	• Input: Limestone (kt/y)	−3036.00
	• Output: Tailing (kt/y)	−1214.40
Aluminum	• Output: Slag (kt/y)	−400.00
Phosphorus chemical	• Output: Phosphorus slag (kt/y)	−500.00
Coal chemical	• Output: Coal gangue (kt/y)	−100.00
Power generation	• Input: Fossil fuel (ktce/y)	−30.00
	• Output: Coal flying ash (kt/y)	−200.00

Guiyang and did have serious negative impacts on the eco-system) reduced by 3164.40 kt/y. In the process consumption, total fossil fuel consumption decreased by 102.37 ktce/y. Solid waste such as slag and flying ash reduced by 1320.00 kt/y. Such benefits were critical to the industrial city, which was engaged in the huge material and energy flows in the life cycles from resource mining, processing, and consumption to waste disposal.

Apart from the direct impacts, the impacts on the whole supply chain were further analyzed with the CFP approach. The total CFP impact was presented in Figure 2.7 by comparison with the scenarios in 2002 and 2012. Results highlight that the implementations of UIS in Guiyang would enable to further reduce the urban CFP by about 1,090,000 tCO_2 annually.

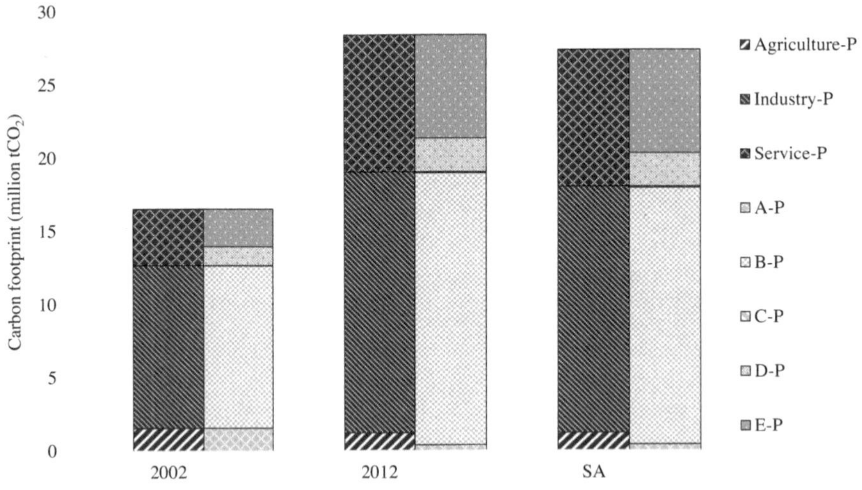

Figure 2.7 Comparison of carbon footprint of UIS scenario and other scenarios

Note: SA represents "Scenario Analysis."

In detail, industrial solid waste exchange, traditional recycling, municipal solid waste utilization, and energy symbiosis would reduce the CFP by 839, 109, 94, and 49 tCO_2/y, respectively. Due to the large amount of exchange, industrial solid waste utilization was still the most effective way to reduce the urban CFP. What's more, since UIS can utilize the waste as raw materials, life cycle benefits can be gained from upstream resource saving. This is the other reason why industrial waste exchange could generate significant CFP mitigation effects. In addition, even though the benefits of municipal solid waste utilization and energy symbiosis were not as large as those of industrial waste use (mainly due to the limited scale), they could be seen as an emerging style of resource and energy reuse, and even the main driver for urban CFP mitigation with the help of more mature and efficient technologies.

2.4.4 Discussion

Based on the review of the state-of-the-art in CFP studies, it has been noticed that so far the mainstream methods for city-level CFP accounting are far from satisfactory. The IPCC method is found incapable of addressing indirect carbon emissions embodied in traded goods, and the hybrid LCA considers both the direct and indirect emissions while neglecting materials beyond food, water, fuels, and cement and still suffering from truncation error in some cases. As a result, an updated IOA method has been employed in this paper that combines the IOTs of Guizhou Province with fiscal data from Guiyang City, both of which are aggregated to the same level of detail.

Contrary to many existing IO-based studies calculating the CFP merely for a single year, we measure the CFPs of 2002, 2007, and 2012 in a comparative sense by taking 2002 as the BAU scenario, in which circular economy promotions have not yet been implemented. The results, therefore, could assist policymakers in tracking the trends in urban CFP but also in understanding the role of a long-term circular economy policy in speeding up urban transition toward a low-carbon society.

Furthermore, we compare the carbon emissions of several cities and provinces as circular economy pilots including Beijing, Liaoning, Shanghai, Shandong, and Chongqing, in comparison to Guiyang (Figure 2.8). As shown below, Guiyang has the smallest amount of CFP. Moreover, it is the only pilot of which CFP began to go down after 2007, which examines that the low-carbon economy and industry transformation related to circular economy strategy has played an important role in carbon emission reduction. Besides, the CFP growth rate of Beijing, Liaoning, and Shandong decreased from 25.1% to 19.3%, 39.6% to 32.3%, and 121.4% to 31.6%, respectively. On the contrary, the CFP growth rate of Chongqing and Shanghai increased from 58.0% to 63.3% and 20.4% to 28.1%, respectively, probably because of their highly advanced industry structure that may shrink the potential in carbon emission reductions. Overall, China's circular economy strategy has proved useful for pilots to mitigate the growing carbon emissions, especially for Guiyang City of which an absolute decoupling of GDP and CFP has achieved since 2007.

However, we acknowledge that our research still exist limitations to be solved in the near future. First, the carbon intensity is calculated with data from Guizhou Province rather than from Guiyang City due to the poor local data availability,

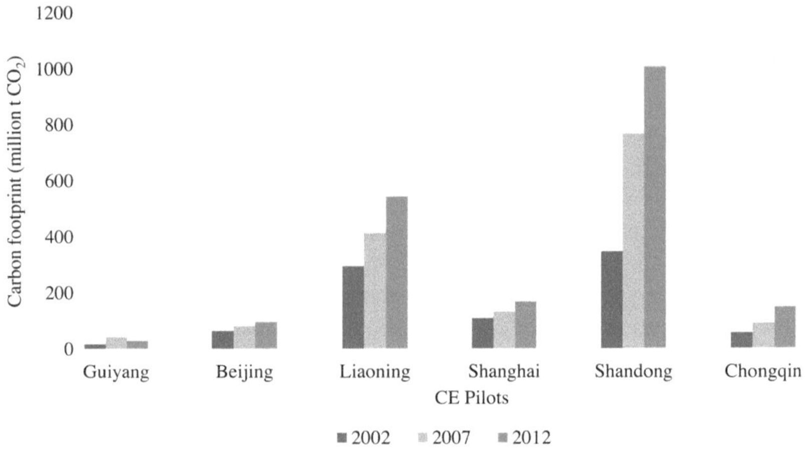

Figure 2.8 Comparison of carbon footprint of circular economy pilots

Note: Data for Guiyang are calculated by this paper, and data for others come from China Emission Accounts & Datasets (CEADs).

which may deviate the final estimate away from the actual condition of the production structure of Guiyang. Second, the inconsistency between the GSY, ESY, and IOTs—the best currently available data sources for Guiyang, may lead to a rough aggregation of sectors. We believe that this issue could be better addressed with the help of increasingly explicit and consistent energy consumption data and fiscal data in addition to the IOTs. Moreover, single-regional IOTs seem not to eliminate the influence of exports when being used to measure the urban CFP from a consumption perspective.

2.5 Conclusions and implications

This paper began with a review of CFP analysis using different methods at multiple scales. On the basis of a deeper understanding of the pros and cons of existing footprinting methods, particularly at the city level, we have made efforts to track the CFP of Guiyang City in three specific years using an updated IOA method by combining the IOTs and energy data of Guizhou Province and the fiscal data of Guiyang City, examine the relationship of CFP between various sectors aggregated to three or five ones, compare the direct and indirect CFPs of production or consumption using two novel indicators named IDRI and PCRI. Moreover, we identified scenarios related to the proposed UIS.

The major findings are as follows: (1) Guiyang's CFP (both in total and per capita) of almost all the sectors increased from 2002 to 2007 and turned to decline afterward as a result of the continuous implementation of circular economy strategy; (2) Industry always took the largest proportion of Guiyang's CFP wherever a production or consumption perspective (especially from a production perspective), suggesting that the supply of Industry to other sectors would be greater than demand; (3) Agriculture had the lowest proportion and the distinction between production-based and consumption-based accounting was insignificant, implying that Agriculture may occupy the middle position in the supply chain mostly; (4) The proportion of Service in consumption-based carbon emissions was much higher than that in production-based carbon emissions, which suggested that Service would be placed in the tail of the supply chain and rely a lot on the inputs from Agriculture or Industry; (5) The CFP of Construction became negligible when being aggregated to five sectors as its proportion in production-based carbon emissions approached zero but, from a consumption perspective, its proportion increased significantly as enormous carbon emissions embodied in the raw materials were taken into account; (6) There was a gap between the allocation of direct and indirect carbon emissions in Construction or Industry, thus adding more stress on the supply chain management; and (7) UIS, in term of industrial solid waste exchange, traditional recycling, municipal solid waste utilization, and energy symbiosis, was able to offer extra CFP mitigation effects via improving the life cycle closing loops.

We hereby drew the conclusion that Guiyang's circular economy promotions had already come into force, particularly after 2007, and would continue to facilitate its urban transition to a low-carbon society. However, we also realized that the present research remained limited. For instance, due to a lack of reliable IOTs at the city level, combining the provincial and urban data was likely to result in a respectively rough aggregation of sectors. Meanwhile, the IOTs from Guizhou Province did not consider the region-of-origin production technology on imported products, thus bringing the risk of misestimating Guiyang's CFP from a consumption perspective. As such, the development of data availability and quality for China's less-developed regions deserve a priority for future endeavors in the modeling and accounting of CFP of urban transition.

Finally, several critical policy implications can be proposed:

1 CFP discloses the differences between direct and indirect carbon emissions based on production and consumption of a given entity, which reflects some hidden linkages among various sectors along the supply chain. By doing so, it could be possible to trace the hot-spot carbon emitters of a region, allowing policymakers to look at the carbon emissions over the life cycle of supply chain with a special focus on sectors and industries that are highly carbon-intensive, but also to launch some incentives or regulations to achieve concrete reduction goals.
2 Attach great importance to the nationwide UIS promotion which, in essence, aims to optimize the supply chain and the relationship of industries and urban area by improving the life cycle closing loops. China owns considerably large-scale industries and the rapid urbanization process, serving as an ideal laboratory to examine the effects of UIS strategy. Therefore, the local stakeholders are encouraged to use this win-win strategy (urban and industrial symbiosis) for economic growth and environmental protection.

3 As a follow-up discussion, it is well known that China has begun to promote the "New Normal Economy Development," which emphasizes the quality of economy rather than the quantity, with more focus on sustainable development and optimized economic structure. As an eco-industrial development measure, UIS does provide an innovative pathway for Guiyang to promote its new normal economic development, via optimizing the interaction of industries and urban areas. For the resource industries in dominant cities like Guiyang, it is necessary to consider how to utilize a systematical approach to better green its industries and; meanwhile, keep economic growth. In addition, innovative decision-making support tools and indicators can help the policymakers to better measure and monitor the new normal economic development by integrating economic, environmental, and ecological values. Our paper verifies the feasibility of CFP as an indicator, which can be considered to be further applied to the index system for new normal economic development evaluation.
4 This paper highlights the significance of R&D to technological innovations in promoting UIS in China. Effective and efficient technologies always provide a fundamental basis for formulating innovative strategies, the key technologies for circular economy and technologies standards, as well as quality standards for the input material as guidance for waste generators.
5 Though CFP has proved useful in dealing with the problems related to emissions, the precondition of accounting for CFP remains challenging, partially due to the lack of access to reliable data, particularly on the meso scales (e.g., cities) and, even if available, the quality of data is often far from satisfactory and varying over time and across regions. Therefore, there is a great need for the establishment of standardized databases (e.g., multi-regional IOTs), which could provide a basis for wise environmental decision-making in various regions.

References

Bai, L., Qiao, Q., Yao, Y., Guo, J., & Xie, M. (2014). Insights on the development progress of national demonstration eco-industrial parks in China. *Journal of Cleaner Production*, *70*, 4–14.

Baldini, C., Gardoni, D., & Guarino, M. (2017). A critical review of the recent evolution of life cycle assessment applied to milk production. *Journal of Cleaner Production*, *140*, 421–435.

Barrett, J., Peters, G., Wiedmann, T., Scott, K., Lenzen, M., Roelich, K., & Le Quéré, C. (2013). Consumption-based GHG emission accounting: A UK case study. *Climate Policy*, *13*(4), 451–470.

Behera, S. K., Kim, J. H., Lee, S. Y., Suh, S., & Park, H. S. (2012). Evolution of 'designed' industrial symbiosis networks in the Ulsan eco-industrial park: 'Research and development into business' as the enabling framework. *Journal of Cleaner Production*, *29*, 103–112.

Bruestle, A. E. (1993). East Asia's urban environment. *Environmental Science & Technology*, *27*(12), 2280–2284.

Chen, M., Liu, W., & Tao, X. (2013). Evolution and assessment on China's urbanization 1960–2010: Under-urbanization or over-urbanization? *Habitat International*, *38*, 25–33.

Chertow, M. R. (2000). Industrial symbiosis: Literature and taxonomy. *Annual Review of Energy and the Environment*, *25*(1), 313–337.

Costa, I., Massard, G., & Agarwal, A. (2010). Waste management policies for industrial symbiosis development: Case studies in European countries. *Journal of Cleaner Production*, *18*(8), 815–822.

Dong, L., Fujita, T., Dai, M., Geng, Y., Ren, J., Fujii, M., Wang, Y., & Ohnishi, S. (2016a). Towards preventative eco-industrial development: An industrial and urban symbiosis case in one typical industrial City in China. *Journal of Cleaner Production*, *114*, 387–400.

Dong, H., Fujita, T., Geng, Y., Dong, L., Ohnishi, S., Sun, L., Dou, Y., & Fujii, M. (2016b). A review on eco-city evaluation methods and highlights for integration. *Ecological Indicators*, *60*, 1184–1191.

Dong, L., Fujita, T., Zhang, H., Dai, M., Fujii, M., Ohnishi, S., Geng, Y., & Liu, Z. (2013a). Promoting low-carbon city through industrial symbiosis: A case in China by applying HPIMO model. *Energy Policy*, *61*, 864–873.

Dong, L., Gu, F., Fujita, T., Hayashi, Y., & Gao, J. (2014b). Uncovering opportunity of low-carbon city promotion with industrial system innovation: Case study on industrial symbiosis projects in China. *Energy Policy*, *65*, 388–397.

Dong, H., Ohnishi, S., Fujita, T., Geng, Y., Fujii, M., & Dong, L. (2014a). Achieving carbon emission reduction through industrial & urban symbiosis: A case of Kawasaki. *Energy*, *64*, 277–286.

Dong, L., Zhang, H., Fujita, T., Ohnishi, S., Li, H., Fujii, M., & Dong, H. (2013b). Environmental and economic gains of industrial symbiosis for Chinese iron/steel industry: Kawasaki's experience and practice in Liuzhou and Jinan. *Journal of Cleaner Production*, *59*, 226–238.

Fang, K., & Heijungs, R. (2015). The role of impact characterization in carbon footprinting. *Frontiers in Ecology and the Environment*, *13*(3), 130–131.

Fang, K., Song, S., Heijungs, R., de Groot, S., Dong, L., Song, J., & Wiloso, E. I. (2016). The footprint's fingerprint: On the classification of the footprint family. *Current Opinion in Environmental Sustainability*, *23*, 54–62.

Feng, K., Davis, S. J., Sun, L., Li, X., Guan, D., Liu, W., Liu, Z., & Hubacek, K. (2013). Outsourcing CO_2 within China. *Proceedings of the National Academy of Sciences*, *110*(28), 11654–11659.

Geng, Y., Tsuyoshi, F., & Chen, X. (2010). Evaluation of innovative municipal solid waste management through urban symbiosis: A case study of Kawasaki. *Journal of Cleaner Production*, *18*(8), 993–1000.

Hamin, E. M., & Gurran, N. (2009). Urban form and climate change: Balancing adaptation and mitigation in the U.S. and Australia. *Habitat International*, *33*, 238–245.

Hashimoto, S., Fujita, T., Geng, Y., & Nagasawa, E. (2010). Realizing CO_2 emission reduction through industrial symbiosis: A cement production case study for Kawasaki. *Resources Conservation and Recycling*, *54*(11), 704–710.

Hillman, T., & Ramaswami, A. (2010). Greenhouse gas emission footprints and energy use benchmarks for eight U.S. cities. *Environmental Science & Technology*, *44*(6), 1902–1910.

Jacobsen, N. B. (2006). Industrial symbiosis in Kalundborg Denmark: A quantitative assessment of economic and environmental aspects. *Journal of Industrial Ecology*, *10*(1–2), 239–255.

Jan, M., Giovanni, B., Thomas, W., John, B., Felix, C., Feng, K., Michael, F., Peter-Paul, P., Helga, W., & Klaus, H. (2013). Carbon footprints of cities and other human settlements in the UK. *Environmental Research Letters*, *8*(3), 035039.

Jiao, W. T., & Boons, F. (2017). Policy durability of circular economy in China: A process analysis of policy translation. *Resources Conservation and Recycling*, *117*, 12–24.
Jukka, H., Mikko, J., Jouni, K. J., Sanna, A.-M., & Seppo, J. (2013). Situated lifestyles: I. How lifestyles change along with the level of urbanization and what the greenhouse gas implications are—a study of Finland. *Environmental Research Letters*, *8*(2), 025003.
Kanemoto, K., Moran, D., & Hertwich, E. G. (2016). Mapping the carbon footprint of nations. *Environmental Science & Technology*, *50*(20), 10512–10517.
Kim, H. W., Ohnishi, S., Fujii, M., Fujita, T., & Park, H. S. (2018). Evaluation and allocation of greenhouse gas reductions in industrial symbiosis. *Journal of Industrial Ecology*, *22*(2), 275–287.
Leontief, W. W. (1936). Quantitative input and output relations in the economic systems of the United States. *The Review of Economics and Statistics*, *18*(3), 105–125.
Li, H., Dong, L., & Ren, J. (2015). Industrial symbiosis as a countermeasure for resource dependent city: A case study of Guiyang China. *Journal of Cleaner Production*, *107*, 252–266.
Liang, S. N., & Jin, Z. H. (2011). Development models of Resource-dependent cities' transformations and its experience and lessons-take Baishan City's development of transformations as an example. *Energy Procedia*, *5*, 1626–1630.
Lin, J. Y., Liu, Y., Meng, F. X., Cui, S. H., & Xu, L. L. (2013). Using hybrid method to evaluate carbon footprint of Xiamen City China. *Energy Policy*, *58*, 220–227.
Liu, Z., Liang, S., Geng, Y., Xue, B., Xi, F., Pan, Y., Zhang, T., & Fujita, T. (2012). Features trajectories and driving forces for energy-related GHG emissions from Chinese mega cites: The case of Beijing Tianjin Shanghai and Chongqing. *Energy*, *37*(1), 245–254.
Liu, H., Wang, X., Yang, J., Zhou, X., & Liu, Y. (2017). The ecological footprint evaluation of low carbon campuses based on life cycle assessment: A case study of Tianjin, China. *Journal of Cleaner Production*, *144*, 266–278.
Liu, H., Zhou, G., Wennersten, R., & Frostell, B. (2014). Analysis of sustainable urban development approaches in China. *Habitat International*, *41*, 24–32.
Meng, X., Wen, Z., Qian, Y., & Yu, H. (2017). Evaluation of cleaner production technology integration for the Chinese herbal medicine industry using carbon flow analysis. *Journal of Cleaner Production*, *163*, 49–57.
Mi, Z., Zhang, Y., Guan, D., Shan, Y., Liu, Z., Cong, R., Yuan, X. -C., & Wei, Y. M. (2016). Consumption-based emission accounting for Chinese cities. *Applied Energy*, *184*, 1073–1081.
Minx, J. C., Wiedmann, T., Wood, R., Peters, G. P., Lenzen, M., Owen, A., Scott, K., Barrett, J., Hubacek, K., Baiocchi, G., Paul, A., Dawkins, E., Briggs, J., Guan, D., Suh, S., & Ackerman, F. (2009). Input-output analysis and carbon footprinting: An overview of applications. *Economic Systems Research*, *21*(3), 187–216.
Mirata, M. (2004). Experiences from early stages of a national industrial symbiosis programme in the UK: Determinants and coordination challenges. *Journal of Cleaner Production*, *12*(8–10), 967–983.
Moriguchi, Y. (2000). Industrial ecology in Japan. *Journal of Industrial Ecology*, *4*(3), 7–9.
NBS (2011). *China Statistical yearbook*. National Bureau of Statistics.
Park, H. S., Rene, E. R., Choi, S. M., & Chiu, A. S. (2008). Strategies for sustainable development of industrial park in Ulsan, South Korea—From spontaneous evolution to systematic expansion of industrial symbiosis. *Journal of Environmental Management*, *87*(1), 1–13.
Ramaswami, A., Chavez, A., Ewing-Thiel, J., & Reeve, K. E. (2011). Two approaches to greenhouse gas emissions foot-printing at the city scale. *Environmental Science & Technology*, *45*(9), 4205–4206.
Ramaswami, A., Hillman, T., Janson, B., Reiner, M., & Thomas, G. (2008). A demand-centered, hybrid life-cycle methodology for city-scale greenhouse gas inventories.

Su, M. R., Chen, B., Xing, T., Chen, C., & Yang, Z. F. (2012). Development of low-carbon city in China: Where will it go? *Procedia Environmental Sciences, 13*, 1143–1148.
Su, B., Huang, H. C., Ang, B. W., & Zhou, P. (2010). Input-output analysis of CO_2 emissions embodied in trade: The effects of sector aggregation. *Energy Economics, 32*(1), 166–175.
Tian, J., Liu, W., Lai, B., Li, X., & Chen, L. (2014). Study of the performance of eco-industrial park development in China. *Journal of Cleaner Production, 64*, 486–494.
Tidy, M., Wang, X., & Hall, M. (2016). The role of supplier relationship management in reducing greenhouse gas emissions from food supply chains: Supplier engagement in the UK supermarket sector. *Journal of Cleaner Production, 112*, 3294–3305.
UN (2012). World urbanization prospects: the 2011 revision. United Nations, Department of Economic and Social Affairs (DESA), Population Division, Population Estimates and Projections Section, New York, 14, 555.
UNFCCC (2015). *Adoption of the Paris Agreement. UNFCCC (United nations framework convention on climate change)*. Proposal by the President. Paris United Nations.
Van Beers, D., Bossilkov, A., Corder, G., & Van Berkel, R. (2007). Industrial symbiosis in The Australian minerals industry: The cases of Kwinana and gladstone. *Journal of Industrial Ecology, 11*(1), 55–72.
Van Berkel, R. (2010). Quantifying sustainability benefits of industrial symbioses. *Journal of Industrial Ecology, 14*(2), 371–373.
Wiedmann, T. (2009). Editorial: Carbon footprint and input-output analysis. *Economic Systems Research, 21*, 175–186.
Wiedmann, T. O., Chen, G., & Barrett, J. (2016). The concept of city carbon maps: A case study of Melbourne, Australia. *Journal of Industrial Ecology, 20*(4), 676–691.
Wu, F. (2012). China's eco-cities. *Geoforum, 43*(2), 169–171.
Xu, M., Weissburg, M., Newell, J. P., & Crittenden, J. C. (2012). Developing a science of infrastructure ecology for sustainable urban systems. *Environmental Science & Technology, 46*, 7928–7929.
Yang, Y., Heijungs, R., & Brandão, M. (2017). Hybrid life cycle assessment (LCA) does not necessarily yield more accurate results than process-based LCA. *Journal of Cleaner Production, 150*, 237–242.
Yang, L., & Li, Y. (2013). Low-carbon city in China. *Sustainable Cities and Society, 9*, 62–66.
Zhang, Q., Nakatani, J., & Moriguchi, Y. (2015). Compilation of an embodied CO_2 emission inventory for China using 135-sector input-output tables. *Sustainability, 7*, 8223.
Zhang, L., Yuan, Z., Bi, J., Zhang, B., & Liu, B. (2010). Eco-industrial parks: National pilot practices in China. *Journal of Cleaner Production, 18*, 504–509.
Zhao, X. L., Cai, Q., Zhang, S. F., & Luo, K. Y. (2017). The substitution of wind power for coal-fired power to realize China's CO_2 emissions reduction targets in 2020 and 2030. *Energy, 120*, 164–178.
Zhou, N., He, G., & Williams, C. (2012). *China's development of low-carbon eco-cities and associated indicator systems (no. LBNL-5873E)*. Lawrence Berkeley National Lab. (LBNL).
Zou, H., Du, H., Wang, Y., Zhao, L., Mao, G., Zuo, J., Liu, Y., Liu, X., & Huisingh, D. (2017). A review of the first twenty-three years of articles published in the journal of cleaner production: With a focus on trends, themes, collaboration networks, low/no-fossil carbon transformations and the future. *Journal of Cleaner Production, 163*, 1–14.

3 Emission Trajectories and Potential Peaking Pathways

3.1 Diverging pathways for sub-nations toward peak emissions

Numerous evidence has demonstrated that global warming caused by anthropogenic greenhouse gases emissions (GHGs) has been one of the major challenges for human well-being (Eyring et al., 2019; Le Quéré et al., 2019; Martinich & Crimmins, 2019). Recently, the Intergovernmental Panel on Climate Change (IPCC, 2014) releases its fifth assessment report that highlights the urgency of controlling GHGs. The 2030 Agenda for sustainable development proposed by the United Nations in 2015 comes up with 17 Sustainable Development Goals (SDGs), where SDG 13 is intended to "take urgent action to combat climate change and its impacts," and SDG 7 to purse "affordable, reliable and sustainable energy." In the same year, the Paris Agreement is reached as a milestone in global climate governance, declaring that the average global temperature increase should stay below 2°C, preferably below 1.5°C (UNFCCC, 2015). Remarkably, it abandons the idea of "top-down" international distributions previously adopted by the Kyoto Protocol and leaves ample room for individual parties to formulate their Nationally Determined Contributions (NDCs) (Fang et al., 2019). As two globally recognized consensuses, how to enhance the synergies between SDGs and the Paris Agreement have received increasing attention, leading to a recent international conference taken place in Copenhagen in April, 2019. In this study, we expect to contribute to aligning NDCs with SDGs.

Recently, global carbon emissions continue increasing and reach a high level of 33431 Mt CO_2 in 2017 without evidence of having peaked yet (BP, 2018). This is essentially because of the rapid urbanization and industrialization in most developing countries, as well as some developed countries. As the world's largest energy consumer and GHGs emitter since 2009 (Wang et al., 2019b), China has formally declared in its NDCs to peak its carbon emissions as early as possible around 2030 (NDRC, 2015). In that sense, whether and how China will reach peak emissions is critical for the fulfillment of not only its NDCs but also the Paris Agreement and the SDGs. Furthermore, investigation into China's peaking paths and time could be informative for other economies, such as India—the country who is supposed to replace China as the biggest carbon emitter in the near future and is unlikely to peak its emissions before 2040 (Jiang et al., 2019a). Thus, focusing on China's peak

DOI: 10.4324/9781003433859-3

emissions can be of global significance. There are quite a few studies on investigation into China's peaking time and emissions at the national level. The results, however, often appear to be confusing or even contradictory (Jiang et al., 2019a). In the case of peaking time, for instance, Guan et al. (2018) assert that China's carbon emissions probably have peaked in 2013, whereas others argue that this could be achieved in 2022–2025 (Yu et al., 2018), 2023–2026 (Song et al., 2018), prior to 2030 (Gallagher et al., 2019), 2030–2035 (Yuan et al., 2014; Zhou et al., 2014), etc. As all this literature differs in many aspects, such as the databases, methodologies, underpinning assumptions, and policy purposes, one may come to the conclusion that predicting peak emissions are entangled with uncertainties that have not yet been resolved; nevertheless, this is a general shortcoming in present correlation and scenario analyses. The problem is that a nationwide analysis may suffer from aggregation bias which overlooks regional heterogeneity (Herrerias et al., 2013; Xu, 2018; Zhang et al., 2018), and therefore can be easy to find in nationwide studies with aggregate data (Brockmeier & Bektasoglu, 2014). Besides, the peaking of China's carbon emissions does not necessarily mean that every Chinese province must peak at the same time, particularly given that the social, economic, and environmental conditions in individual provinces vary remarkably. As such, there is a great need for revisiting China's emissions trajectory and peaking time at the provincial level, and the estimates in aggregate can somehow overcome the aggregation bias and thus be of national significance. In this way, the estimates are likely to be more precise and reliable, with policy implications that are differentiated and implementable for regional energy saving and climate change mitigation. To our understanding, however, there is a lack of research into peak emissions that takes into account all the Chinese provinces simultaneously.

Several approaches have been adopted to understand the historical and future trajectories of carbon emissions at multiple scales, such as the grey model, environmental Kuznets Curves (EKC), and Stochastic Impacts by Regression on Population, Affluence, and Technology (STIRPAT). The grey model roots from the grey system theory aimed at finding the statistical relationship between various social, economic, technical, and other variables (Deng, 1982). By means of the grey model, it is predicted that China will peak its energy consumption and associated carbon emissions in 2014 and 2020, respectively (Pao et al., 2012). In a more recent estimation, researchers suggest that China may reach peak emissions by 2030 (Li et al., 2018). However, grey analysis has been criticized for its inherent limitations in long-term prediction, the lack of transparency in methodology, and the incapability of discerning driving forces (Wu et al., 2015; Zeng et al., 2018).

The EKC analysis has a much wider spectrum of applicability than the grey model as it provides an important tool to analyze how carbon emissions dynamically evolve with economic development (Yin et al., 2015). It comes from an empirical hypothesis that there is an inverse U-shaped relationship between environmental indicators and economic growth (Grossman and Kruger, 1991). Several studies examine the turning points in China's carbon emissions per capita with rising income levels by using time series or panel data (Alam et al., 2016; Yuan et al., 2014). However, EKC researches are often criticized for considering few factors and equating turning points with peaks (Beckerman, 1992). Some scholars report that the turning points outside are likely to coincide with semi-log or log-log curves instead of invert-U curves (Liddle, 2015; Richmond & Kaufmann, 2006).

The IPAT equation is intended to quantify the impacts of population (P), affluence (A), and technology (T) on the environment (Ehrlich & Holdren, 1971). Nonetheless, this model is flawed because of its failure to track non-linear relationship between them and the exclusion of additional driving forces of specific environmental indicators (Wang et al., 2017). To overcome these demerits, STIRPAT is created with the aim of extending and testing various formulations of the IPAT hypothesis (Dietz & Rosa, 1997; Ehrlich & Holdren, 1971). The STIRPAT analysis has now been recognized as an effective way of identifying the social, economic, and technical driving factors behind energy-related carbon emissions (Zhang and Lin, 2012; Yao et al., 2015).

For instance, by combining an extended STIRPAT model with a system dynamics model, Liu and Xiao (2018) observe that China's carbon emissions are to peak either in 2023, in 2025, or unlikely under three scenarios. This brings the subject of scenario analysis into focus, which has been recognized as complementary to STIRPAT analysis (Wu et al., 2018b). Different scenario analyses have divergent applicability and limitations in short-, medium-, and long-term forecasts at multiple scales (Grubler et al., 2018; O'Neill et al., 2017). Anyhow, to obtain a scientifically robust prediction of emissions trajectories and peaking paths, a scenario analysis is often integrated into a STIRPAT model.

Research into peak emissions in China focuses primarily on the nation as a unit of analysis, and minorly on few provinces (Jiang et al., 2019b). For instance, Liu and Xiao (2018) assemble population, GDP per capita, energy intensity, energy structure, total fixed-asset investment, and industrial structure in an extended STIRPAT model with ten scenarios, and the results show that China is likely to peak emissions around 2023–2025 and others have no obvious decline trends. Meanwhile, few scholars have brought single provinces or cities into focus. By using the extended STIRPAT model to analyze the factors of carbon emissions with the existing scenarios to predict carbon emissions, Beijing will peak before 2030 and achieve its intensity target before 2020 (Liu et al., 2018). In another case, an extended STIRPAT model is used to determine the relationship between carbon emissions and seven driving factors (i.e., population, GDP per capita, energy intensity, urbanization, energy consumption structure, the ratio of service industry, and foreign trade degree) with eight scenarios in Qingdao city, and the results show that peak emissions will be reached in 2020 under three scenarios and in 2025 under the rest scenarios (Wu et al., 2018a).

In principle, the STIRPAT analysis represents a step ahead from classical EKC in the sense that it takes into account more factors affecting carbon emissions, to predict potential pathways to reach peak emissions in a sound way. Aligning STIRPAT models with scenario analysis would be appropriate for short-term and medium-term prediction (Du et al., 2012), especially useful for decision makers who are committed to seeking for feasible emission reduction paths since all the possible outcomes can be observed in a comparative sense (Lin & Du, 2017; Wang et al., 2019a).

To sum up, the possible peak-reaching time and peak emissions of Chinese provinces remain largely unexplored in existing literature. To close this major gap in our knowledge, this paper attempts to employ time series data for each province to investigate the driving forces behind the energy-related carbon emissions in 30 Chinese provinces (excluding Taiwan), explore whether and when each province will peak its emissions under different scenarios, and identify the optimal peaking paths for different provinces. By doing so, this paper is able to make novel contributions to the literature on three aspects: (1) shifting the focus from the whole nation to every province, which is the key to making sense of China's potential peaking paths while having been largely neglected in research into peak emissions; (2) going beyond looking at peaking time merely and paying more attention to cumulative carbon emissions in determining optimal peaking paths for each province, particularly given its close relation to global warming; and (3) bringing clarity to the bottom-up approach aimed to support better prediction and evaluation of possible emissions trajectory for other nations and the whole world and provide solution to aggregation bias due to the gap between provincial data in aggregate and the nationwide data.

3.2 Methods and data

To begin with, we account for the energy-related CO_2 emissions (all the emissions mentioned in this paper are expressed in CO_2 Mt, although in the following the unit is abbreviated as Mt and we name CO_2 emissions as carbon emissions for reasons of simplicity) of 30 provinces between 1995 and 2015, and explore the driving forces such as industrial structure, urbanization, energy structure, etc., by developing a set of extended STIRPAT models on the basis of ridge regression, and different scenarios for projecting the emissions trajectory by 2045 by province. Furthermore, we identify the best scenarios for all the provinces and explore potential peaking paths that China as a whole may achieve.

3.2.1 Energy-related carbon emissions accounting

We account for the energy-related carbon emissions associated with coal, oil, natural gas, and non-fossil energy by using the method recommended by IPCC (2006) in the following equation:

$$C_p = \frac{44}{12}\sum_{i=1}^{4}(F_{p,i}E_{p,i}) \tag{3.1}$$

where C_p is the carbon emissions of province p; and $F_{p,i}$ and $E_{p,i}$ are the carbon emission factor of energy of type i and consumption of the energy of type i for province p, respectively. The energy consumption of four types is converted into standard coal consumption according to their respective heat values (Yang et al., 2018). In our research, the energy-related carbon emission coefficients are obtained from the Chinese Energy Research Institute (Wu et al., 2018a).

3.2.2 *STIRPAT analysis*

STIRPAT in essence is a stochastic regression model:

$$I = aP^b A^c T^d e, \tag{3.2}$$

where I is a given environmental indicator; a is a constant term of this model; P, A, and T are population, affluence, and technology, respectively; b, c, and d are the exponential parameters to be estimated; and e is the random error term. Equation (3.2) could be transformed into logarithms as follows:

$$lnI = lna + b(lnP) + c(lnA) + d(lnT) + lne \tag{3.3}$$

The STIRPAT model could be extended by integrating additional factors, including population (P), GDP per capita, the quadratic term of GDP per capita, energy intensity (EI), urbanization rate (UR), energy structure (ES), and industrial structure (IS), following the equation:

$$\begin{aligned} lnC = {} & lna + blnP + cln(GDP/P) + dln(GDP/P)^2 + elnEI \\ & + flnUR + glnES + hlnIS + lne \end{aligned} \tag{3.4}$$

where C is carbon emissions. The ordinary least square (OLS) regression is originally employed for parameter estimation as per Equation (3.4), but it fails to capture the relationship between provincial carbon emissions and those socioeconomic factors due to the multicollinearity of these independent variables. In view of this, ridge regression, as proposed by Hoerl and Kennard (1970), comes into focus because it can improve the accuracy and reliability of parameter estimation through bias-variance trade-offs (Li et al., 2018; Liddle, 2013; Wang et al., 2014a).

3.2.3 *Scenario analysis*

Potentially, long-term carbon emissions of each province are extremely sensitive to the variation in a suite of socioeconomic factors in the future. We attempt to take advantage of scenario analysis to capture those possible emissions trajectories. Here we identify three scenarios: low-speed, business as usual (BAU), and high-speed, and each of them is aggregated to four time periods, namely 2016–2020, 2021–2025, 2026–2030, and post-2030.

In the BAU scenarios, the annual rates of changes in factors during 2016–2020 are set in accordance with the 13th Five-Year Plan of each province, with slightly decreasing trends for most of the factors. Compared with the BAU scenarios, the annual rates of change in factors are set somewhat higher or lower under the high-speed and low-speed scenarios, respectively, depending on the models established and factors included. Table 3.1 illustrates the three scenarios for the case of Chongqing. As per Chongqing's 13th Five-Year Plan, the urbanization rate in Chongqing is expected to be 65% by 2020 and the population growth rate will range from 3.62‰ to 6.50‰. Therefore, the annual growth of population and urbanization in Chongqing are set at

Table 3.1 Scenario setup for Chongqing

Scenario	*Period*	*P*	*UR*	*GDP/P*	*IS*	*EI*	*ES*
Low	2016–2020	0.40%	5.00%	5.50%	–1.60%	–6.00%	–1.80%
	2021–2025	0.35%	4.80%	5.00%	–1.40%	–5.00%	–1.60%
	2026–2030	0.30%	4.60%	4.50%	–1.20%	–4.00%	–1.40%
	Post-2030	0.25%	4.40%	4.00%	–1.00%	–3.00%	–1.40%
BAU	2016–2020	0.50%	6.00%	6.00%	–2.20%	–7.00%	–2.00%
	2021–2025	0.45%	5.50%	5.50%	–2.00%	–6.00%	–1.90%
	2026–2030	0.40%	5.00%	5.00%	–1.80%	–5.00%	–1.80%
	Post-2030	0.35%	4.50%	4.50%	–1.60%	–4.00%	–1.70%
High	2016–2020	0.70%	7.00%	9.00%	–2.30%	–9.00%	–2.40%
	2021–2025	0.60%	6.50%	7.00%	–2.20%	–8.00%	–2.30%
	2026–2030	0.55%	6.00%	6.00%	–2.10%	–7.00%	–2.20%
	Post-2030	0.50%	5.50%	5.50%	–2.00%	–6.00%	–2.00%

Note: *P, UR, GDP/P, IS, EI*, and *ES* are population, urbanization rate, GDP per capita, industrial structure, energy intensity, and energy structure, respectively.

a speed of 5‰ and 6% in 2016–2020, respectively. GDP would grow at a rate of 7% over the same period. Together with the trends of GDP and population, we further set a growth rate of 6% for GDP per capita. The variations in industrial structure, energy intensity, and energy structure are set to be at –2.2%, –7%, and –2%, respectively. For other periods of time, the annual rates of change in these factors are adjusted according to the historical data of each province.

3.2.4 *Data and variables*

As seen in Table 3.2, we select six main factors in constituting our STIRPAT models. In most cases, population serves as a major driving factors of carbon emissions (Yue et al., 2013). GDP per capita as a main indicator of affluence leads to increase in carbon emissions as well (Lin & Xie, 2014). Energy intensity is form of technology index that facilitates mitigating climate change (Shahbaz et al., 2015). The contributions of other drivers like industrial structure, urbanization rate, and energy structure have been examined in numerous researches, even though the results appear to be discrepant or even contradictory (Lin et al., 2016; Liu & Xiao, 2018; Wang et al., 2019a).

The data for various driving forces by province are extracted from provincial statistical yearbooks (1996–2016). The data for energy consumption by province are derived from China Energy Statistical Yearbook and China Statistical Yearbook (1996–2016).

3.3 Results

3.3.1 *Socioeconomic drivers of carbon emissions for provinces*

We first carry out a correlation analysis of each pair of the variables including carbon emissions, population, per capita GDP, energy intensity, industrial structure, urbanization rate, and energy structure, observing a significant correlation between

Table 3.2 Data sources and definitions of driving forces

Variable	*Definition*	*Sources*
Carbon emissions	The amount of carbon emissions from fuel energy consumption	China Energy Statistical Yearbook and provincial statistical yearbooks (1996–2016)
Population	Total population	China Statistical Yearbook (1996–2016)
GDP per capita	Total GDP divided by end-year population	China Statistical Yearbook and provincial statistical yearbooks (1996–2016)
Energy intensity	The amount of energy consumed to produce a certain level of GDP	China Energy Statistical Yearbook and provincial statistical yearbooks (1996–2016)
Industrial structure	Ratio of secondary industry added value in GDP	Provincial statistical yearbooks (1996–2016)
Urbanization	Share of residents living in urban areas in total population	Provincial statistical yearbooks (1996–2016), China Compendium of Statistics and Provincial Socioeconomic Statistics Bulletin (1996–2016)
Energy structure	Proportion of coal fuel in total energy consumption	China Energy Statistical Yearbook (1996–2016)

most of the factors. Furthermore, by applying OLS regression, we notice that in many cases the variance inflation factors (VIFs) for individual provinces are over 10, and some even over 100. We eliminate the multicollinearity across variables in almost every province using ridge regression. While population, urbanization, GDP per capita, industrial structure, energy structure, and energy intensity have been selected as influencing factors for all the 30 provinces, in some cases those factors that do not pass the significance test of the ridge regression with 90-percentile confidence intervals have been excluded from the empirical STIRPAT models. The resulting regression coefficients of these socioeconomic factors by province can be found in Table 3.3.

Increase in population, ratio of secondary industry, energy intensity, and proportion of coal fuels would contribute to accelerated carbon emissions for all the provinces. For example, every 1% increase in Chongqing's population is likely to lead to an increase of 5.40% carbon emissions. When it comes to Shanxi, however, the carbon emissions would only increase by 0.05%. It is surprising that in some provinces (e.g., Anhui, Henan, Guangxi, Qinghai) the variations in population have no significant impact on carbon emissions.

The carbon emissions of Guizhou are quite sensitive to the variation in industrial structure, as proved by the fact that every 1% increment in the ratio of secondary industry would result in an increase in carbon emissions by 1.69%. By contrast, the same improvement in industrial structure in Shanghai can trigger an increase of 0.01% carbon emissions. In the meanwhile, the carbon emissions of ten provinces, such as Liaoning, Heilongjiang, Zhejiang, are decoupling from the variation

Table 3.3 Results of ridge regression and STIRPAT analyses for 30 provinces

Province	*lnP*	*ln(GDP/P)*	*ln(GDP/P)*^{2}	*lnUR*	*lnEI*	*lnIS*	*lnES*	*Cons*	*K*	R^2
Anhui	1.059	0.149***	0.006***	0.330**	0.267***	0.101	0.393	−0.985*	0.10	0.93
Beijing	1.611*	0.232***	0.006***	0.758*	0.028*	1.019*	0.417*	−8.802***	0.20	0.91
Chongqing	5.402***	0.323***	−0.132***	0.776***	0.218***	0.132*	0.144***	−2.956**	0.04	0.95
Fujian	2.910*	0.223***	−0.018***	0.118***	0.313**	0.414	0.840**	−6.900***	0.15	0.95
Gansu	1.087***	0.217***	0.012***	1.059***	0.646**	1.258***	0.793	5.841	0.20	0.96
Guangdong	0.454***	0.102***	−0.013***	0.005**	0.334***	0.199*	0.268***	4.021***	0.30	0.92
Guangxi	0.632	0.166***	0.009***	0.500***	0.309***	0.529	0.214	13.200*	0.30	0.92
Guizhou	0.992***	0.354***	0.019***	0.068	0.093	1.690***	1.009***	1.962	0.15	0.95
Hainan	0.571*	0.546***	0.314 ***	1.023***	0.014	0.568*	1.210*	2.960***	0.10	0.86
Hebei	2.046**	0.187	0.008	0.461***	0.045*	1.029*	0.045	−8.577	0.05	0.96
Heilongjiang	1.661*	0.153***	0.048***	0.142*	0.089	0.053	0.035	−6.574*	0.10	0.87
Henan	2.616	0.221***	0.012***	0.424***	0.121	0.007	1.516**	−5.707	0.10	0.92
Hubei	1.885**	1.032**	−0.310***	0.268**	0.191**	0.424*	0.139***	−6.649***	0.05	0.84
Hunan	3.094***	0.171*	0.009***	0.649***	0.707	0.075	1.294**	6.699**	0.15	0.92
Inner Mongolia	0.278*	0.234**	0.012***	0.234**	1.009*	0.032*	0.117	−2.932	0.10	0.94
Jiangsu	2.672***	0.135***	−0.181***	0.205***	0.007*	0.015***	0.699	−5.455**	0.20	0.95
Jiangxi	0.087***	0.131***	0.072***	0.279***	0.225***	0.442**	0.025	8.915***	0.20	0.96
Jilin	1.542*	0.122***	0.209***	−0.043*	0.188**	1.605*	0.247*	−9.951*	0.15	0.92
Liaoning	3.800**	0.114***	0.006***	0.437***	0.263	0.162	0.569	−5.882***	0.10	0.90
Ningxia	2.520***	0.186***	0.009***	0.79***	0.273	0.086	0.391	−9.215***	0.10	0.90
Qinghai	0.745	0.247***	−0.013***	0.881**	0.181	0.088	0.368**	0.780	0.10	0.91
Shaanxi	1.131***	0.353***	0.018***	−0.153*	0.001	1.346***	0.358	−3.474**	0.05	0.97
Shandong	0.095***	0.143***	0.078***	0.282***	0.232***	0.312*	0.096	8.716***	0.1	0.97
Shanghai	0.381*	0.250*	−0.059***	−0.246*	0.642**	0.009**	1.573***	6.724***	0.2	0.8
Shanxi	0.047*	0.321***	0.011**	0.574*	0.047*	0.038	1.336**	−10.581	0.02	0.9
Sichuan	0.914**	0.345***	0.014***	0.254	0.059	1.130**	1.068**	−0.624	0.05	0.97
Tianjin	0.327	0.271***	−0.109***	0.669*	1.017***	0.292**	0.571**	6.904***	0.05	0.98
Xinjiang	0.879**	0.214***	0.011***	1.076**	0.077	0.75**	0.238	3.659	0.1	0.92
Yunnan	0.814***	0.395***	0.021***	0.501***	0.072	1.017***	1.189*	−0.459	0.32	0.96
Zhejiang	1.109**	0.189***	0.008***	0.301*	0.272***	0.496	0.611**	−1.668	0.2	0.92

Note: *lnP*, $ln\left(\frac{GDP}{P}\right)$, $ln\left(\frac{GDP}{P}\right)^2$, *lnUR*, *lnEI*, *lnIS*, and *lnES* are logarithmic population, GDP per capita and its quadratic form, urbanization rate, energy intensity, industrial structure, and energy structure, respectively. ***, **, and * represent $p<0.01$, $p<0.05$, and $p<0.1$, respectively.

of industrial structure. Likewise, no significant correlation is observed between energy intensity and structure and carbon emissions for approximately half provinces. With every 1% increase in energy intensity, the carbon emissions of Tianjin and Shanghai increase by 1.02% and 0.64%, respectively.

Urbanization in general would drive provincial carbon emissions, with exceptions of Shanghai, Shaanxi, and Jilin, where every 1% increase in urbanization rate agrees with decrease in carbon emissions by 0.25%, 0.15%, and 0.04%, respectively. The reasons are complex, of which one might be that rural-urban migration makes people make use of cleaner energy (Shen et al., 2017). Moreover, we find it interesting that the carbon emissions of Sichuan and Guizhou are insensitive to urbanization (Zhang & Lin, 2012; Zhou & Liu, 2016).

Generally, the growth of GDP per capita would lead to more carbon emissions for all the provinces. The fastest increase in emissions is found in Hubei, followed by Yunnan, Guizhou, and Shaanxi. As GDP per capita continues rising, its contribution to the carbon emissions of nine provinces (e.g., Hubei, Jiangsu, Chongqing) reverses, becoming an inhibitory factor instead.

3.3.2 Peak-reaching time and emissions of provinces

Figure 3.1 displays the provincial carbon emissions in each scenario from 2015 to 2045. For majority of the 30 provinces, the cumulative carbon emissions in high scenarios are higher than in low and BAU scenarios prior to the peak and, conversely, the carbon emissions in high scenarios are often lower than in low and BAU scenarios after reaching the peak emissions. Examples include Inner Mongolia, Zhejiang, Hubei, etc. Moreover, the peaking time for 11 provinces (e.g., Xinjiang and Guangxi) is postponed under high scenarios, whereas ten provinces (e.g., Guizhou and Anhui) would peak earlier under high scenarios. One should be noted that for a province the carbon emissions in low scenarios are not necessarily larger than those in high and BAU scenarios, as both the positive and negative factors differ simultaneously among the low, BAU, and high scenarios. This can be interpreted as evidence of the complexity of potential emissions trajectory and peaking path driven by diverse factors that reinforce or counteract each other.

In accordance with the different peak-reaching time in these three scenarios, we divide 30 provinces into four groups: (1) 12 provinces (Beijing, Tianjin, Shanghai, Xinjiang, Gansu, Qinghai, Hubei, Liaoning, Guangdong, Chongqing, Jiangsu, and Zhejiang) are likely to peak their carbon emissions before 2030 regardless of the scenarios, and particularly Shanghai and Xinjiang could probably peak the emissions in 2019 under low scenarios; (2) Five provinces are likely to peak emissions prior to 2030 in one or two scenarios. Of these, Fujian and Shanxi would peak in low and BAU scenarios, Hainan in BAU and high scenarios, Anhui in high scenario, and Yunnan in BAU scenario; (3) Nine provinces (Inner Mongolia, Shaanxi, Jiangxi, Hunan, Henan, Sichuan, Guizhou, Guangxi, and Shandong) would peak emissions between 2031 and 2043 in one or more scenarios; (4) Hebei, Ningxia, Liaoning, and Heilongjiang are unlikely to peak emissions by 2045 in any recognizable scenarios.

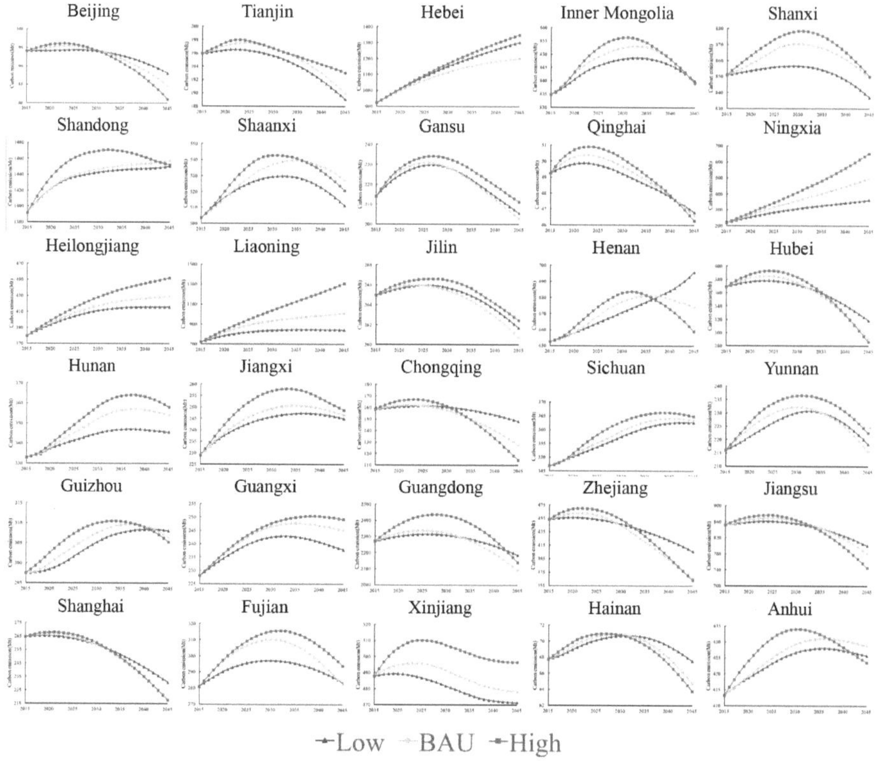

Figure 3.1 Energy-related carbon emissions of 30 provinces in low, BAU, and high scenarios

In low-speed scenarios, eight provinces (e.g., Shanghai, Xinjiang, Zhejiang) would peak emissions in the period 2021–2025, and six provinces (e.g., Guangdong, Beijing, Jilin) in 2026–2030, for which the average peak emissions are 373.80 Mt (Figure 3.2). However, the magnitude of peak emissions varies dramatically. Qinghai and Jiangsu, for example, would peak their emissions at 49.88 Mt in 2022 and at 861.32 Mt in 2024, respectively. Meanwhile, ten provinces (e.g., Hainan, Shaanxi, Yunnan) would reach peak emissions between 2031 and 2045. The remaining six provinces are incapable of peaking emissions before 2045, such as Hebei, Ningxia, and Liaoning.

In BAU scenarios, 14 provinces can hardly peak their carbon emissions before 2030 (Figure 3.3). Specifically, the earliest peak emissions might be observed for Shanghai, with an estimate of 266.87 Mt in 2020. 12 provinces would reach peaks between 2021 and 2030, where the least peak emissions (50.36 Mt) are found in Qinghai in 2023 and, conversely, the largest peak emissions (870.35 Mt) are found in Shanxi in 2030. Nine provinces, including Inner Mongolia, Shaanxi, Jiangxi, etc., would have their peak emissions in the period 2033–2040, with an average value of 448.15 Mt. Five provinces (e.g., Hebei, Heilongjiang, Liaoning) are incapable of peaking their emissions before 2045, mostly located in North and Northeast China.

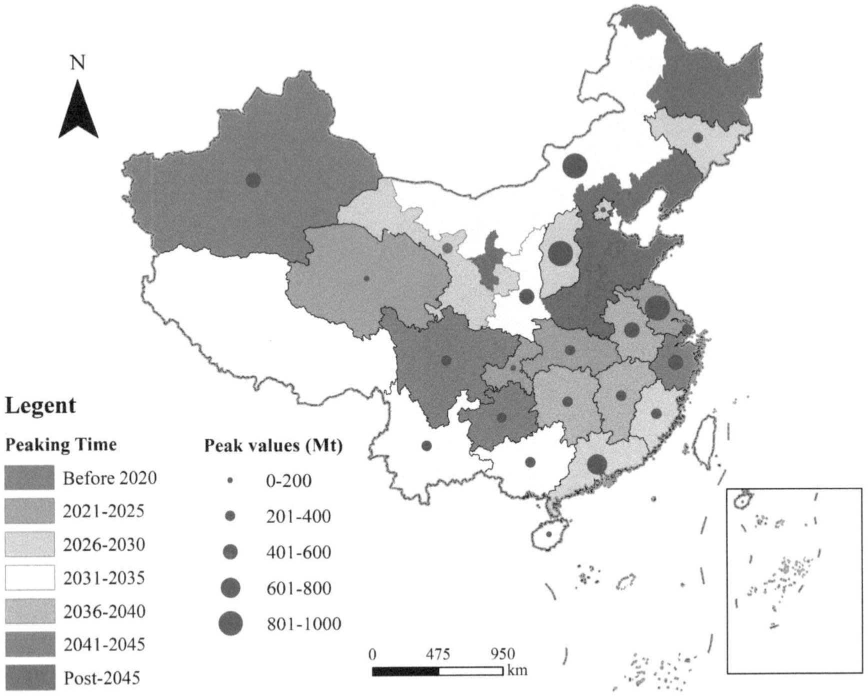

Figure 3.2 Peaking time for 30 provinces in low scenarios

Unlike the BAU scenarios, the high-speed scenarios allow only 14 provinces (e.g., Shanghai, Zhejiang, Chongqing) to peak emissions before 2030 (Figure 3.4). The least and largest peak emissions are observed for Qinghai and Jiangsu, with an estimate of 50.89 Mt in 2023 and 876.31 Mt in 2024, respectively. Similarly, Shanghai remains the earliest peak-reaching time, with an estimate of 267.99 Mt in 2021. 12 provinces, such as Shanxi, Yunnan, and Inner Mongolia, would peak their emissions between 2031 and 2039 at 544.36 Mt on average. The remaining four provinces, including Hebei, Ningxia, Liaoning, and Heilongjiang, fail to reach peak emissions prior to 2045.

3.3.3 Optimal peaking paths for provinces

While China has promised to peak its carbon emissions no later than 2030, concrete evidence of whether and how to fulfill this ambitious goal at the provincial level is lacking. On the other hand, we argue that focusing exclusively on peak-reaching time runs the risk of enlarging cumulative emissions, which is defined in this paper as the amount of total carbon emissions of a region in the period 2016–2045. Essentially, the consensus-based 2°C target is subject to the cumulative emissions rather than to the peaking time. Besides, seeking the peaking time

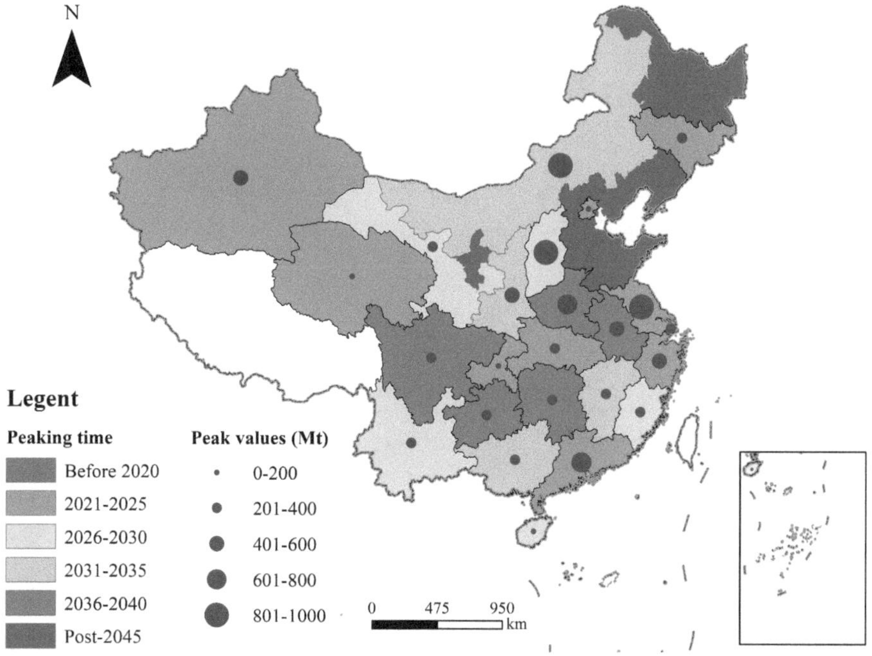

Figure 3.3 Peaking time for 30 provinces in BAU scenarios

as soon as possible is an action in order to control the cumulative carbon emissions which is our terminal target. As such, it makes great sense to seek for the optimal peaking paths for each of the 30 provinces. To that end, we divide these 30 provinces into four categories on the basis of the preceding analysis.

1 12 provinces can peak before 2030 in all the three scenarios. For provinces of this category, cumulative emissions till the peaking time would play a central important role in determining the so-called optimal peaking paths, particularly given the consensus on maintaining global temperature increase within 2°C. In other words, the most appropriate scenario setup for these provinces must ensure the smallest cumulative carbon emissions. For this concern, we consider low-speed scenarios as the optimal paths for Xinjiang, Gansu, Tianjin, and Qinghai, BAU scenarios for Hubei, Liaoning, Guangdong, Chongqing, Shanghai, Zhejiang, and Beijing, and high-speed scenario for Jiangsu.
2 Five provinces cannot reach peaks before 2030 in at least one scenario. For provinces of this category, peaking time and cumulative emissions are considered simultaneously in selecting optimal paths, with the intent of allowing for peaking prior to 2030 with relatively small cumulative carbon emissions. In this case, we find that BAU scenarios are a good fit for Shanxi, Yunan, and Hainan, high scenario for Anhui, and low scenario for Fujian.

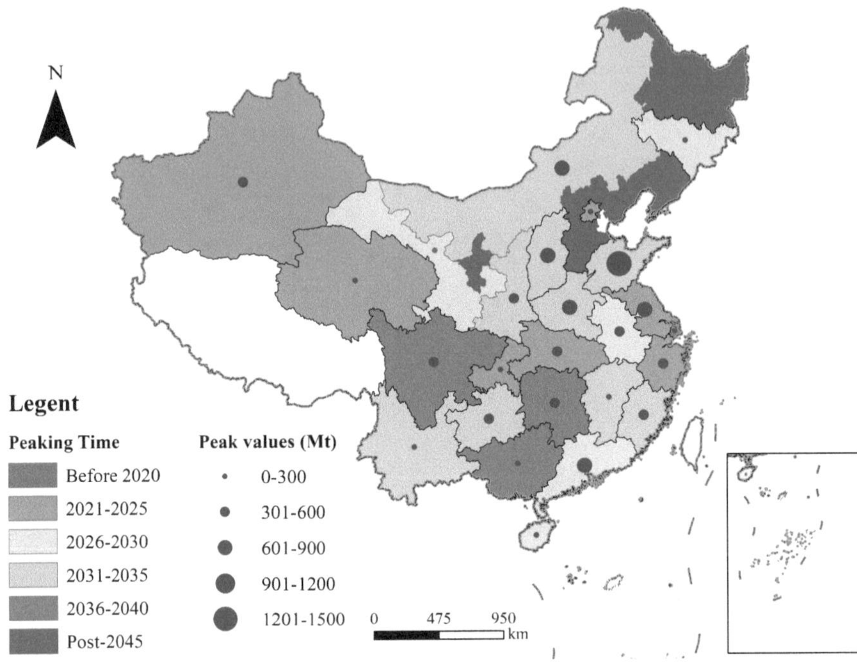

Figure 3.4 Peaking time for 30 provinces in high scenarios

3 Nine provinces are likely to reach peaks between 2031 and 2045. For provinces of this category, peaking time is the major concern when choosing optimal paths. If two scenarios hold the same peaking time, the scenario with less cumulative emissions would be preferred. As a result, high-speed scenarios are suitable for Inner Mongolia, Shaanxi, Jiangxi, Henan, Sichuan, Guizhou, and Shandong and, conversely, low-speed scenarios are suitable for Hunan and Guangxi.
4 Last but not least, four provinces cannot peak in any of the three scenarios by 2045. This is largely because of the fact that some of the factors that often contribute to emissions reduction, such as energy intensity, industrial structure, and energy structure, may have been excluded from the STIRPAT analyses of these provinces as they do not pass the significance test of the regression. For provinces of this category, scenarios in keeping with the smallest cumulative emissions till 2045 should be selected as the optimal paths. For this reason, low scenarios appear to be relatively appropriate for Hebei, Ningxia, Liaoning, and Heilongjiang.

Figure 3.5 displays the possible peaking time and cumulative emissions for the 30 provinces under low, BAU, and high scenarios. By contrasting these provinces into the four categories, we make it possible to identify the optimal pathways for each province, as illustrated in Table 3.4.

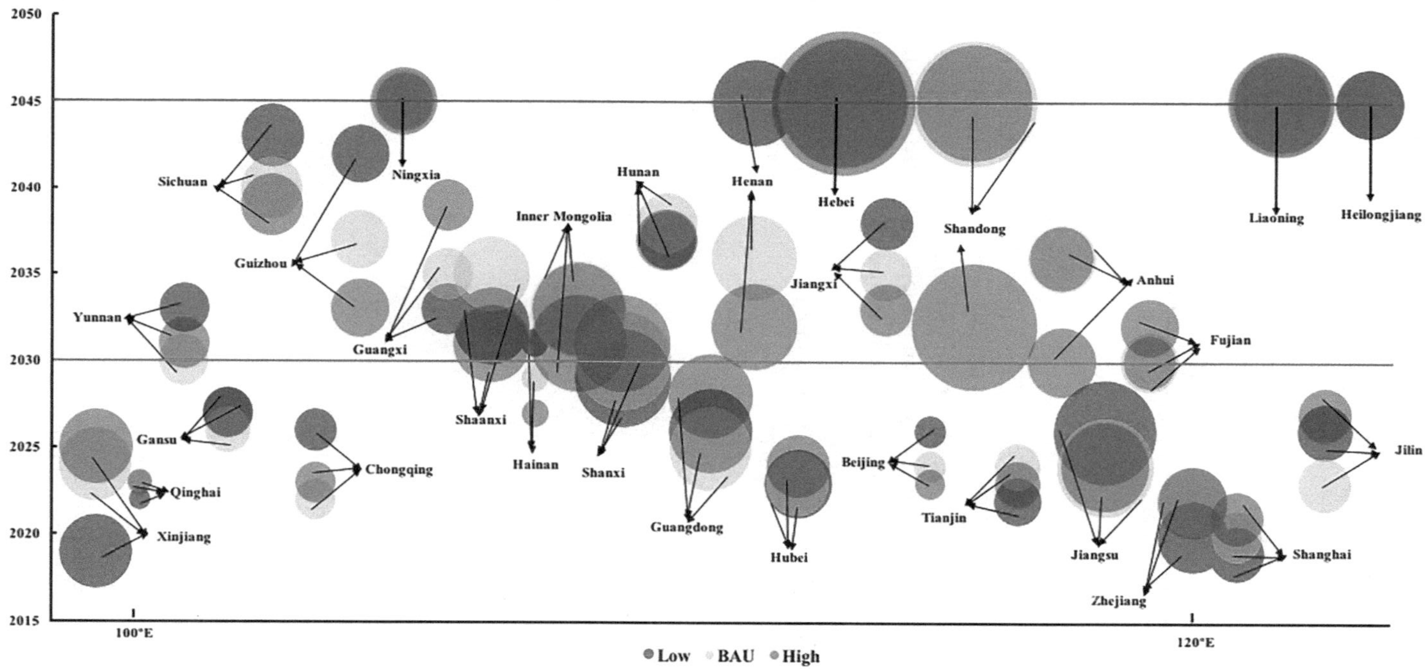

Figure 3.5 Cumulative carbon emissions of 30 provinces between 2015 and 2045

Note: The provinces are arranged by longitude of their own capital cities in ascending order. The bubble size represents the amount of cumulative carbon emissions. The light gray, medium gray and dark gray represent the situation in low, BAU, and high scenarios, respectively. The center height of the bubbles represents the peaking time. The upper line represents peaking in 2030, and the bottom line represents peaking in or after 2045.

Table 3.4 The optimal scenarios for 30 provinces by considering peaking time and cumulative carbon emissions

	Low scenarios	*BAU scenarios*	*High scenarios*
Provinces	Xinjiang, Hebei, Gansu, Fujian, Qinghai, Tianjin, Ningxia, Liaoning, Heilongjiang, Guangxi, Hunan	Shanghai, Chongqing, Zhejiang, Hainan, Jilin, Beijing, Hubei, Guangdong, Shanxi, Yunnan	Shandong, Jiangsu, Anhui, Inner Mongolia, Shaanxi, Guizhou, Henan, Jiangxi, Sichuan

3.4 Discussion

3.4.1 Comparison with literature

Prediction of the emissions trajectory of Chinese individual provinces, particularly the peak emissions, has been a growing area of concern. A comparison of our analysis and existing few studies that anticipate the peaking time for China and seven Chinese provinces based on different regression models is provided in Figure 3.6(a) and (b). Our estimates are generally consistent with others, as witnessed by the

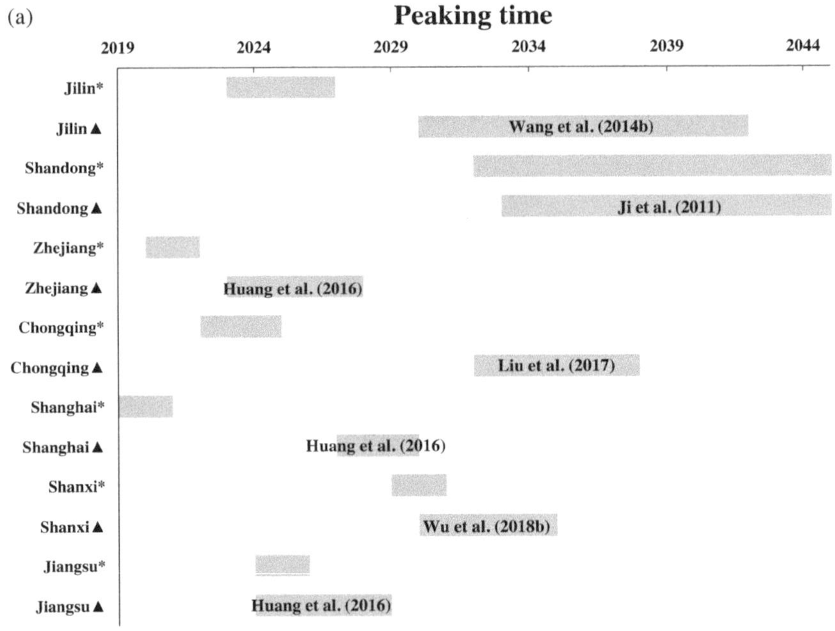

Figure 3.6 Comparison of our results with the literature on the anticipated peak emissions in single provinces (a) and China as a whole (b) (*Continued*)

Note: In Figure 3.6(a), * represents the results from this paper, and ▲ represents the results from other papers. In Figure 3.6(b), we use three stars to show the peaking time and peak emissions. The left one represents the path in which China will peak by earliest in 2028, the median one represents the optimal path in which China will peak in 2030 with the lowest cumulative emissions of 403607 Mt, and the right one represents the path in which China will peak by latest in 2040.

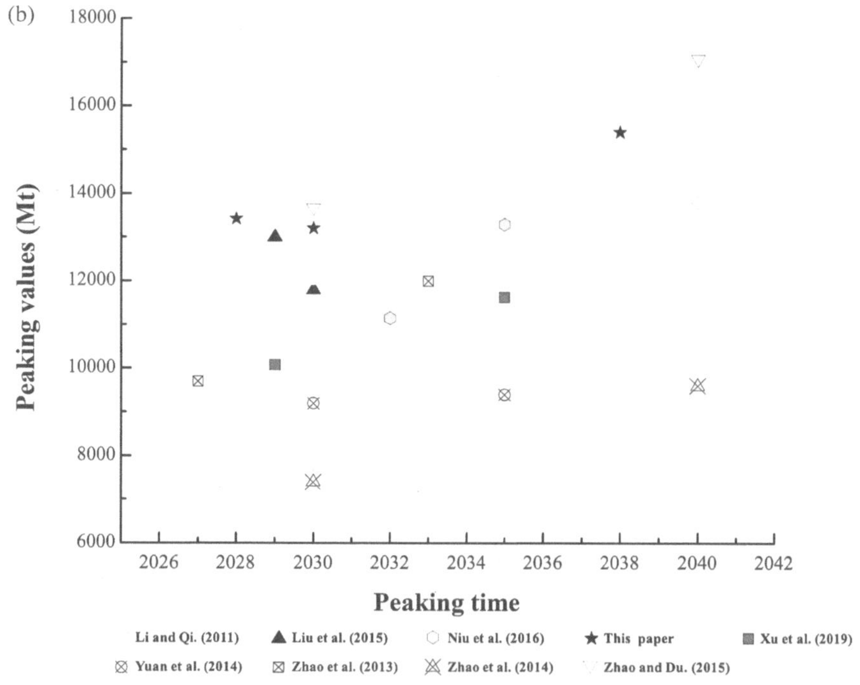

Figure 3.6 (Continued)

cases of Shandong, Shanxi, and Gansu. As for the other four provinces (Jilin, Zhejiang, Chongqing, Shanghai), we find it easier than previously thought to reach peak emissions. The most notable gaps between our estimates and others are observed in Shanghai and Chongqing, for which we believe that the earliest peaking time could be 2019 and 2022, respectively. Because of the growing evidence that China will peak its emissions well in advance of 2030 (Gallagher et al., 2019; Mi et al., 2017; Yu et al., 2018) and the fact that some developed provinces are expected to peak emissions around 2020 (Huang et al., 2017; NDRC, 2015), these two years seem more reasonable than those of 2027 and 2032 anticipated by Liu et al. (2018) and Li et al. (2011) for Shanghai and Chongqing, respectively.

The main reason of the discrepancy in results between this paper and others is that we make use of a consistent set of extended STIRPAT models that assemble different factors selected by ridge regression analysis, with the aim of ensuring the comparability, consistency, and reliability of our estimation. By contrast, the results of other studies follow from different types of quantitative analysis. Yet, one should be noted that research of this nature inevitably involves large uncertainties; therefore, basically a province may reach its peak emissions within a period of years rather than in a specific year. As illustrated in Figure 3.6(a), the uncertainty of peaking time for the selected seven provinces ranges from 2 years up to 13 years, depending on how many drivers are taken into account under specific scenarios. We consider this uncertainty range of peaking time to be more rational than proclaiming peak emissions in a specific single year.

While a variety of disparate methods and models have been adopted to forecast the peaking of carbon emissions (Niu et al., 2016), most studies conclude that China will peak its emissions between 2025 and 2040, with peak values ranging from 8000 Mt to 18000 Mt, as illustrated in Figure 3.6(b). In our study, we start with focusing on the provincial scale to allow for more accurate estimation by province and better inform region-specific policy making. Our provincial results in aggregate can somehow overcome the aggregation bias resulting from nationwide data, demonstrating that China is highly likely to reach peak emissions between 2028 and 2040 (Figure 3.7). The estimates are in well agreement with previous studies presented in Figure 3.6(b). Specifically, we find that China would peak by earliest in 2028, for which the peak and cumulative emissions will be 13425 Mt and 412356 Mt, respectively. By contrast, the latest peaking year for China is estimated to be 2040, by which the peak and cumulative emissions will be 15040 Mt and 442811 Mt, respectively. Furthermore, we see that peaking in 2030 might be optimal for China, as in this way the country would have the cumulative emissions of 403607 Mt, which is lower than those in any other potential paths. Nevertheless, apparently not all the provinces will peak by no later than 2030. Examples include Hebei, Inner Mongolia, Shaanxi, Liaoning, Heilongjiang, Jiangxi, Shandong, Henan, Hunan, Guangxi, Sichuan, and Guizhou.

3.4.2 Policy implications

Trends in research into peak emissions are now in danger of focusing exclusively on the national scale and overlooking the concrete evidence of potential emissions trajectory by province. However, there are strong arguments for understanding climate within the rapidly changing climate-sensitive systems in the developing countries through bottom-up approaches that are proved effective in modeling simulations of the future (Conway et al., 2019). Our study, in that sense, is able to provide a basis

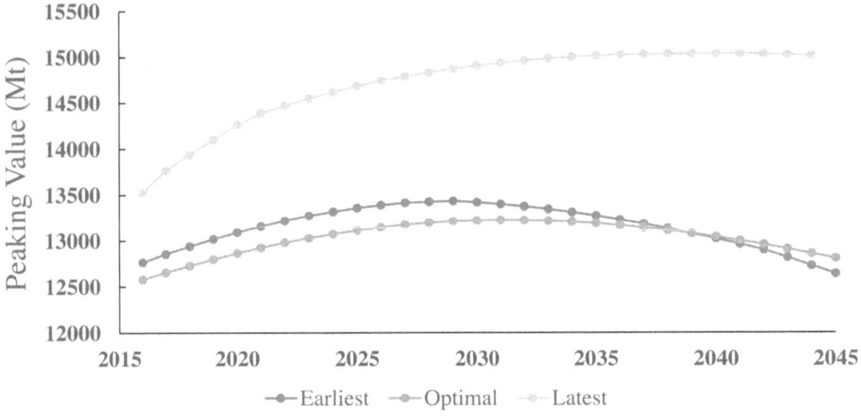

Figure 3.7 The earliest, optimal, and latest peaking paths for China overall

for assigning China's NDCs and informing "common but differentiated" energy and climate policies on the provincial scale where climate policy actions most commonly take place, but also for yielding a scientifically robust nationwide analysis by integrating the estimates of peak emissions by province. We propose the following policy recommendations: (1) Provinces that will always peak before 2030 should give priority to monitoring the growth of population (e.g., for Shanghai, Chongqing, and Jiangsu) and urbanization (e.g., for Beijing, Tianjin, and Chongqing); (2) For provinces that may peak before 2030 in one or two scenarios, like Hainan, Fujian, and Shaanxi, technological progress is found to be the key to advancing peaking time and lowering peak emissions; (3) For provinces unlikely to peak before 2030, it is necessary to keep their population growth, urbanization, and the changes in energy structure at rates compatible with optimal scenarios; and (4) Provinces that are unlikely to peak in any recognizable scenarios by 2045 must implement new policies to introduce more low-carbon technologies, encourage more environmental-friendly consumption patterns, and formulate stricter environmental regulations to combat climate change (Miao et al., 2019; Yin et al., 2015).

The upscaling of estimates on the provincial paths to peak emissions in a suite of scenarios demonstrates that China as a whole will peak its emissions between 2028 and 2040, with cumulative emissions ranging from 403607 Mt (corresponding to peaking in 2030) to 456191 Mt (corresponding to peaking in 2038). It would be preferred if China can reach the peak in 2030, as in this case its cumulative emissions will be in a minimum value, even though still not all the provinces will peak by 2030. This illustrates a very important point; that is, the peaking of China's carbon emissions does not guarantee that every Chinese province must peak at the same time. More importantly, our findings imply that focusing exclusively on the peak-reaching time may lead to unexpected increase in cumulative emissions and therefore compromise the pursuit of China's NDCs. Because of the near-linear relationship between cumulative emissions and global warming (IPCC, 2014), we argue for going beyond looking at peaking time merely and paying more attention to cumulative carbon emissions in determining optimal peaking paths for both China and its individual provinces. To that end, cumulative emissions should be gained prominence in the evaluation of provincial government performance on implementing NDCs. Rather than carrying on the never-ending debate over the exact time when China peaks its emissions at the national level, we are convinced that it makes more sense to go to great lengths to analyze the emissions trajectory by province, especially if the goal is to show the geographical heterogeneity of fulfilling China's NDCs and to inform regional policy making. Provincial results in aggregate have the potential to yield reasonable and implementable policy recommendations for seeking high-quality paths to peak. This could not only contribute to achieving the Paris Agreement but also facilitate the alignment of NDCs with the endorsement of a number of SDGs, in particular SDG 13 on climate action and SDG 7 on energy. The pursuit of SDG 7, for instance, has been found imperative to most of the 169 targets in the 2030 Agenda for Sustainable Development (Nerini et al., 2018). Additionally, there is a need for China to road-map new energy and climate strategies in the post-2030 timeframe, which is lacking in its present NDCs and SDGs (Guan & Delman, 2017).

3.4.3 Uncertainties and limitations

As said, any research on the future emissions trajectories and peaking paths inevitably contains uncertainties and limitations. Our paper trying to take a preliminary investigation into China's peak emissions at the provincial level is not an exception. First, while the carbon emissions from energy consumption activities are mainly derived from the burning of fossil energy, there might be additional sources of emissions that do not included in the official statistics. Our emissions accounting, in that sense, serves as an incomplete inventory analysis. In contrast to tracing the flows of emissions embodied in traded products and services by use of multi-regional input–output analysis (Fang et al., 2014; Long et al., 2019), we think that it makes more sense to provide a real picture of territorial energy-related emissions for different Chinese provinces when exploring the peaking paths by province.

Second, the impacts of different variables on the energy-related carbon emissions of provinces can be quantified in a number of ways. Although OLS is an ordinary method to conduct such a regression analysis, it may be not appropriate for use in the context of severe multicollinearity, as proved in our case where most variance inflation factors appear too large to be acceptable. By contrast, ridge regression has argued to be a good fit for running an extended STIRPAT model by substantially reducing multicollinearity (Wang et al., 2013). The danger of multicollinearity lies in the generation of large standard errors among related independent variables that makes the regression analysis unstable (Wang et al., 2013). Ridge regression provides a cost-effective solution to the multicollinearity problem by doing an acceptably biased estimate with smaller mean square errors through bias-variance trade-offs.

Third, uncertainty is incorporated into the data and scenario setting: (1) Despite the authoritative sources of data in this paper, the transparency, reliability, and accuracy of these data have been steeped in controversy, which introduces some uncertainties to the calculations; (2) Six widely adopted variables have been employed to understand the drivers of energy-related carbon emissions by province, while the appropriateness of this selection remains to be seen. It might be desirable to cover a wide-enough range of factors, so more data such as those of fixed assets investment are supposed to collect in support of the STIRPAT analysis. But this perhaps can add uncertainty to the final estimates if the data are far from satisfactory; (3) While scenarios can function as appropriate tools for evaluating the impacts of driving forces on future emissions, as is done in myriad literature including IPCC series reports, such an analysis is primarily reliant on historical data and expert knowledge that inevitably contain subjectivity and arbitrariness; and (4) Future trends of variables are filled with uncertainty as well. As a result, our prediction is indicative and sensitive to policy decisions and might somehow move the final estimates away from the true values. More scenario settings could be introduced to capture a broader range of variation in diverse driving factors.

3.5 Conclusions

Starting with accounting for the energy-related carbon emissions of the 30 Chinese provinces, we develop a set of extended STIRPAT models assembling time series data to investigate the driving factors behind the variation in carbon emissions, predict the emissions trajectory in different scenarios and identify the optimal peaking paths by province, and finally take a bottom-up approach to explore whether and when China will peak its emissions. Our analysis indicates that the earliest peak-reaching scenarios may not be the best option for those provinces whose cumulative carbon emissions are estimated to be considerably large, and that peaking time and cumulative emissions should hold the same importance in determining the optimal peaking paths both on the national and provincial scales. Provincial results in aggregate explicitly show that China would reach peak emissions in the period 2028–2040, preferably in 2030 to allow for the lowest cumulative carbon emissions of the country. Conversely, the highest cumulative emissions are observed for the path to peak in 2038. To that end, we suggest that: (1) For provinces that will always peak before 2030, the BAU scenarios are the best fit for Shanghai, Chongqing, Zhejiang, Jilin, Beijing, Tianjin, Hubei, and Guangdong, the low scenarios are for Gansu, Qinghai, and Xinjiang, and the high scenario is for Jiangsu; (2) For provinces that may peak before 2030 in one or two scenarios, the low, BAU, and high scenarios are preferred for Fujian, for Yunnan and Hainan, and for Anhui, and Shanxi, respectively; (3) For provinces unlikely to peak before 2030, the low and high scenarios are more appropriate for Hunan and Guangxi, and for Jiangxi, Shaanxi, Henan, Shandong, Inner Mongolia, Sichuan, and Guizhou, respectively; and (4) The rest four provinces (Hebei, Ningxia, Liaoning, and Heilongjiang) who are unlikely to peak in any recognizable scenarios must take immediate action to improve the economic, social, and demographic policies.

From a broader point of view, this study can provide decision makers with the following policy advices on implementation and development of China's NDCs: (1) Reasonable emission reduction commitments cannot been assigned without in-depth discussions of the potential peaking paths in well-designed scenarios on the provincial scale, which is largely neglected by the academic and policy communities; (2) To avoid unexpected increase in carbon emissions in the future, a shift of focus from looking merely at the peak-reaching time of a region to considering both peaking time and the associated cumulative emissions is needed; and (3) China should road-map its follow-up strategies for energy conservation and climate governance in the post-2030 timeframe, such as building net-zero emissions energy systems, in order to fill in the gap in its present NDCs and SDGs. Furthermore, we are convinced that our findings are informative for anticipating possible emissions trajectory in other nations and the whole world, and therefore would be of global significance.

References

Alam, M. M., Murad, M. W., Noman, A. H. M., & Ozturk, I. (2016). Relationships among carbon emissions, economic growth, energy consumption and population growth: Testing environmental Kuznets curve hypothesis for Brazil, China, India and Indonesia. *Ecological Indicators*, *70*, 466–479.

Beckerman, W. (1992). Economic growth and the environment: Whose growth? Whose environment? *World Development*, *20*(4), 481–496.

BP. (2018). BP Energy Outlook 2018 edition - data pack. Available at: http://www.bp.com/energyoutlook

Brockmeier, M., & Bektasoglu, B. (2014). Model structure or data aggregation level: Which leads to greater bias of results? *Economic Modelling*, *38*, 238–245.

Conway, D., Nicholls, R. J., Brown, S., Tebboth, M. G., Adger, W. N., Ahmad, B., Biemans, H., Crick, F., Lutz, A. F., De Campos, R. S., Said, M., Singh, C., Ludi, E., Zaroug, M. A. H., New, M., & Wester, P. (2019). The need for bottom-up assessments of climate risks and adaptation in climate-sensitive regions. *Nature Climate Change*, *9*, 503–511.

Deng, J. L. (1982). Control problems of grey systems. *Systems & Control Letters*, *1*(5), 288–294.

Dietz, T., & Rosa, E. A. (1997). Effects of population and affluence on CO_2 emissions. *Proceedings of the National Academy of Sciences of the United States of America*, *94*(1), 175–179.

Du, L., Wei, C., & Cai, S. (2012). Economic development and carbon dioxide emissions in China: Provincial panel data analysis. *China Economic Review*, *23*(2), 371–384.

Ehrlich, P. R., & Holdren, J. P. (1971). Impact of population growth. *Science*, *171*(3977), 1212–1217.

Eyring, V., Cox, P. M., Flato, G. M., Gleckler, P. J., Abramowitz, G., Caldwell, P., Collins, W. D., Gier, B. K., Hall, A. D., Hoffman, F. M., Hurtt, G. C., Jahn, A., Jones, C. D., Klein, S. A., Krasting, J. P., Kwiatkowski, L., Lorenz, R., Maloney, E., Meehl, G. A., Pendergrass, A. G., Pincus, R., Ruane, A. C., Russell, J. L., Sanderson, B., Santer, M., Sherwood, B. D., Simpson, S. C., Stouffer, I. R., & Williamson, R. J. (2019). Taking climate model evaluation to the next level. *Nature Climate Change*, *9*, 102–110.

Fang, K., Heijungs, R., & de Snoo, G. R. (2014). Theoretical exploration for the combination of the ecological, energy, carbon, and water footprints: Overview of a footprint family. *Ecological Indicators*, *36*, 508–518.

Fang, K., Zhang, Q., Long, Y., Yoshida, Y., Sun, L., Zhang, H., Dou, Y., & Li, S. (2019). How can China achieve its intended nationally determined contributions by 2030? A multi-criteria allocation of China's carbon emission allowance. *Applied Energy*, *241*, 380–389.

Gallagher, K. S., Zhang, F., Orvis, R., Rissman, J., & Liu, Q. (2019). Assessing the policy gaps for achieving China's climate targets in the Paris Agreement. *Nature Communications*, *10*, 1256.

Grossman, G. M., & Kruger, A. B. (1991). Environmental impacts of the North American free trade agreement. NBER Working Paper. No. 3914.

Grubler, A., Wilson, C., Bento, N., Boza-Kiss, B., Krey, V., McCollum, D. L., Rao, N. D., Riahi, K., Rogelj, J., Stercke, S. D., Cullen, J., Frank, S., Fricko, O., Guo, F., Gidden, M., Havlík, P., Huppmann, D., Kiesewetter, G., Rafaj, P., Schoepp, W., & Valin, H. (2018). A low energy demand scenario for meeting the 1.5°C target and sustainable development goals without negative emission technologies. *Nature Energy*, *3*, 515–527.

Guan, T., & Delman, J. (2017). Energy policy design and China's local climate governance: Energy efficiency and renewable Energy policies in Hangzhou. *Journal of Chinese Governance*, *2*(1), 68–90.

Guan, D., Meng, J., Reiner, D. M., Zhang, N., Shan, Y., Mi, Z., Shao, S., Liu, Z., Zhang, Q., & Davis, S. J. (2018). Structural decline in China's CO_2 emissions through transitions in industry and energy systems. *Nature Geoscience*, *11*, 551–555.

Herrerias, M. J., Joyeux, R., & Girardin, E. (2013). Short-and long-run causality between energy consumption and economic growth: Evidence across regions in China. *Applied Energy*, *112*, 1483–1492.

Hoerl, A. E., & Kennard, R. W. (1970). Ridge Regression: Biased estimation for nonorthogonal problems. *Technometrics*, *12*(1), 55–67.

Huang, R., Lu, Y., & Lu, M. (2017). Projection of energy consumption carbon emission peak for Jiangsu, Zhejiang and Shanghai under different energy policies. *Resources and Environment in the Yangtze Basin*, *26*(1), 15–25.

IPCC (2006). *2006 IPCC guidelines for national greenhouse gas inventories*. IGES Press.

IPCC (2014). *Climate change 2014: Synthesis report*. Cambridge University Press.

Jiang, J., Ye, B., & Liu, J. (2019a). Research on the peak of CO_2 emissions in the developing world: Current progress and future prospect. *Applied Energy*, *235*, 186–203.

Jiang, J., Ye, B., & Liu, J. (2019b). Peak of CO_2 emissions in various sectors and provinces of China: Recent progress and avenues for further research. *Renewable and Sustainable Energy Reviews*, *112*, 813–833.

Ji, J., Luan, S., & Jiang, X. (2011). Carbon emissions analysis and prediction of shandong province based on the modified STIRPAT model. *Chinese Journal of Management Science*, *19*, 687–689.

Le Quéré, C., Korsbakken, J. I., Wilson, C., Tosun, J., Andrew, R., Andres, R. J., Canadell, J. G., Jordan, A., Peters, G. P., & van Vuuren, D. P. (2019). Drivers of declining CO_2 emissions in 18 developed economies. *Nature Climate Change*, *9*, 213–217.

Liddle, B. (2013). Urban density and climate change: A STIRPAT analysis using City-level data. *Journal of Transport Geography*, *28*, 22–29.

Liddle, B. (2015). What are the carbon emissions elasticities for income and population? Bridging STIRPAT and EKC via robust heterogeneous panel estimates. *Global Environmental Change*, *31*, 62–73.

Li, H., Mu, H., Zhang, M., & Li, N. (2011). Analysis on influence factors of China's CO_2 emissions based on Path–STIRPAT model. *Energy Policy*, *39*(11), 6906–6911.

Lin, B., & Du, Z. (2017). Promoting energy conservation in China's metallurgy industry. *Energy Policy*, *104*, 285–294.

Lin, B., Omoju, O. E., Nwakeze, N. M., Okonkwo, J. U., & Megbowon, E. T. (2016). Is the environmental Kuznets curve hypothesis a sound basis for environmental policy in Africa? *Journal of Cleaner Production*, *133*, 712–724.

Lin, B., & Xie, C. (2014). Reduction potential of CO_2 emissions in China's transport industry. *Renewable and Sustainable Energy Reviews*, *33*, 689–700.

Li, H., & Qi, Y. (2011). Comparison of China's carbon emission scenarios in 2050. *Advances in Climate Change Research*, *2*(4), 193–202.

Liu, Q., Li, Q., & Zheng, X. (2017). The prediction of carbon dioxide emissions in Chongqing based on fossil fuel combustion. *Acta Scienciae Circumstantiae*, *37*(4), 1582–1593.

Liu, D., & Xiao, B. (2018). Can China achieve its carbon emission peaking? A scenario analysis based on STIRPAT and system dynamics model. *Ecological Indicators*, *93*, 647–657.

Liu, J., Zhang, X., & Song, X. (2018). Regional carbon emission evolution mechanism and its prediction approach driven by carbon trading–A case study of Beijing. *Journal of Cleaner Production*, *172*, 2793–2810.

Li, F., Xu, Z., & Ma, H. (2018). Can China achieve its CO_2 emissions peak by 2030? *Ecological Indicators*, *84*, 337–344.

Long, Y., Yoshida, Y., Fang, K., Zhang, H., & Dhondt, M. (2019). City-level household carbon footprint from purchaser point of view by a modified input-output model. *Applied Energy*, *236*, 379–387.

Martinich, J., & Crimmins, A. (2019). Climate damages and adaptation potential across diverse sectors of the United States. *Nature Climate Change*, *9*, 397–404.

Miao, Z., Baležentis, T., Tian, Z., Shao, S., Geng, Y., & Wu, R. (2019). Environmental performance and regulation effect of China's atmospheric pollutant emissions: Evidence from "three regions and ten urban agglomerations. *Environmental and Resource Economics*, *74*(1), 211–242.

Mi, Z., Wei, Y. M., Wang, B., Meng, J., Liu, Z., Shan, Y., Liu, J., & Guan, D. (2017). Socioeconomic impact assessment of China's CO_2 emissions peak prior to 2030. *Journal of Cleaner Production*, *142*, 2227–2236.

National Development Reform Commission (NDRC) (2015). Enhanced Actions on Climate Change. Available at http://www4.unfccc.int/ndcregistry/PublishedDocuments/China%20First/China%27s%20First%20NDC%20Submission.pdf.

Nerini, F. F., Tomei, J., To, L. S., Bisaga, I., Parikh, P., Black, M., Borrion, A., Spataru, C., Broto, V. C., Anandarajah, Milligan, B. G., & Mulugetta, Y. (2018). Mapping synergies and trade-offs between energy and the sustainable development goals. *Nature Energy*, *3*, 10–15.

Niu, S., Liu, Y., Ding, Y., & Qu, W. (2016). China's energy systems transformation and emissions peak. *Renewable and Sustainable Energy Reviews*, *58*, 782–795.

O'Neill, B. C., Kriegler, E., Ebi, K. L., Kemp-Benedict, E., Riahi, K., Rothman, D. S., van Ruijven, B. J., van Vuuren, D. P., Birkmann, J., Kok, K., Levy, M., & Solecki, W. (2017). The roads ahead: Narratives for shared socioeconomic pathways describing world futures in the 21st century. *Global Environmental Change*, *42*, 169–180.

Pao, H. T., Fu, H. C., & Tseng, C. L. (2012). Forecasting of CO_2 emissions, energy consumption and economic growth in China using an improved grey model. *Energy*, *40*(1), 400–409.

Richmond, A. K., & Kaufmann, R. (2006). Is there a turning point in the relationship between income and energy use and/or carbon emissions? *Ecological Economics*, *56*(2), 176–189.

Shahbaz, M., Nasreen, S., Abbas, F., & Anis, O. (2015). Does foreign direct investment impede environmental quality in high-, middle-, and low-income countries? *Energy Economics*, *51*, 275–287.

Shen, H., Tao, S., Chen, Y., Ciais, P., Güneralp, B., Ru, M., Zhong, Q., Yun, X., Zhu, X., Huang, T., Tao, W., Chen, Y., Li, B., Wang, X., Liu, W., Liu, J., & Zhao, S. (2017). Urbanization-induced population migration has reduced ambient $PM_{2.5}$ concentrations in China. *Science Advances*, *3*(7), e1700300.

Song, J., Yang, W., Wang, S., Wang, X. E., Higano, Y., & Fang, K. (2018). Exploring potential pathways towards fossil energy-related GHG emission peak prior to 2030 for China: An integrated input-output simulation model. *Journal of Cleaner Production*, *178*, 688–702.

United Nations Framework Convention on Climate Change (UNFCCC) (2015). Adoption of the Paris Agreement. Report No. FCCC/CP/2015/L.9/Rev.1. Avaiable at http://unfccc.int/resource/docs/2015/cop21/eng/l09r01.pdf.

Wang, S., Fang, C., Guan, X., Pang, B., & Ma, H. (2014a). Urbanisation, energy consumption, and carbon dioxide emissions in China: A panel data analysis of China's provinces. *Applied Energy*, *136*, 738–749.

Wang, Z., Huang, W., & Chen, Z. (2019a). The peak of CO_2 emissions in China: A new approach using survival models. *Energy Economics*, *81*, 1099–1108.

Wang, X., Wang, Y., & Duan, H. (2014b). Forecasting area's carbon emissions of energy consumption and controllability study. *China Population Resources & Environment*, *24*(8), 9–16.

Wang, S., Wang, J., Li, S., Fang, C., & Feng, K. (2019b). Socioeconomic driving forces and scenario simulation of CO_2 emissions for a fast-developing region in China. *Journal of Cleaner Production*, *216*, 217–229.

Wang, C., Wang, F., Zhang, X., Yang, Y., Su, Y., Ye, Y., & Zhang, H. (2017). Examining the driving factors of energy related carbon emissions using the extended STIRPAT model based on IPAT identity in Xinjiang. *Renewable and Sustainable Energy Reviews*, *67*, 51–61.

Wang, P., Wu, W., Zhu, B., & Wei, Y. (2013). Examining the impact factors of energy-related CO_2 emissions using the STIRPAT model in Guangdong province, China. *Applied Energy*, *106*, 65–71.

Wu, C., Huang, G., Xin, B., & Chen, J. (2018a). Scenario analysis of carbon emissions' anti-driving effect on Qingdao's energy structure adjustment with an optimization model, part I: Carbon emissions peak value prediction. *Journal of Cleaner Production*, *172*, 466–474.

Wu, L., Liu, S., Liu, D., Fang, Z., & Xu, H. (2015). Modelling and forecasting CO_2 emissions in the BRICS (Brazil, Russia, India, China, and South Africa) countries using a novel multi-variable grey model. *Energy*, *79*, 489–495.

Wu, L., Wang, J., & Guo, P. (2018b). Peak regional carbon emissions based on open STIRPAT modeling in an energy-producing region of Shanxi. *Resources Science*, *40*(5), 1051–1062.

Xu, T. (2018). Investigating environmental Kuznets curve in China–aggregation bias and policy implications. *Energy Policy*, *114*, 315–322.

Yang, L., Xia, H., Zhang, X., & Yuan, S. (2018). What matters for carbon emissions in regional sectors? A China study of extended STIRPAT model. *Journal of Cleaner Production*, *180*, 595–602.

Yao, C., Feng, K., & Hubacek, K. (2015). Driving forces of CO_2 emissions in the G20 countries: An index decomposition analysis from 1971 to 2010. *Ecological Informatics*, *26*, 93–100.

Yin, J., Zheng, M., & Chen, J. (2015). The effects of environmental regulation and technical progress on CO_2 Kuznets curve: An evidence from China. *Energy Policy*, *77*, 97–108.

Yuan, J., Xu, Y., Hu, Z., Zhao, C., Xiong, M., & Guo, J. (2014). Peak energy consumption and CO_2 emissions in China. *Energy Policy*, *68*, 508–523.

Yue, T., Long, R., Chen, H., & Zhao, X. (2013). The optimal CO_2 emissions reduction path in Jiangsu province: An expanded IPAT approach. *Applied Energy*, *112*, 1510–1517.

Yu, S., Zheng, S., Li, X., & Li, L. (2018). China can peak its energy-related carbon emissions before 2025: Evidence from industry restructuring. *Energy Economics*, *73*, 91–107.

Zeng, B., Duan, H., Bai, Y., & Meng, W. (2018). Forecasting the output of shale gas in China using an unbiased grey model and weakening buffer operator. *Energy*, *151*, 238–249.

Zhang, Q., Liao, H., & Hao, Y. (2018). Does one path fit all? An empirical study on the relationship between energy consumption and economic development for individual Chinese provinces. *Energy*, *150*, 527–543.

Zhang, C., & Lin, Y. (2012). Panel estimation for urbanization, energy consumption and CO_2 emissions: A regional analysis in China. *Energy Policy*, *49*, 488–498.

Zhao, X., & Du, D. (2015). Forecasting carbon dioxide emissions. *Journal of Environmental Management*, *160*, 39–44.

Zhao, Z., Wang, R., Voss, H., & Yan, Y. (2013). The prediction of China's carbon emission turning point based on the classic EKC model. *Finance & Trade Economics*, *34*(10), 81–88.

Zhou, Y., Hao, F., Meng, W., & Fu, J. (2014). Scenario analysis of energy-based low-carbon development in China. *Journal of Environmental Sciences*, *26*(8), 1631–1640.

Zhou, Y., & Liu, Y. (2016). Does population have a larger impact on carbon dioxide emissions than income? Evidence from a cross-regional panel analysis in China. *Applied Energy*, *180*, 800–809.

4 Key Sectors in Achieving Peak Emissions

4.1 Relation between industrial and national emissions

The climate change associated with carbon emissions poses unprecedented threats to the international community, such as extreme weather, sea-level rise, infectious diseases, biodiversity loss, and food shortage (Mora et al., 2018). To mitigate climate warming, a consensus was reached worldwide that end-of-century warming must be maintained below 2°C, or even 1.5°C above the pre-industrial levels (McGlade & Ekins, 2015; Tong et al., 2019; UNFCCC, 2015). To this end, a growing number of nations have recently committed to realizing net-zero emission systems that do not increase the atmospheric carbon concentration by 2050 (IPCC, 2018). However, global emissions continue, reaching a record-high level of 33.3 Gt in 2019 (IEA, 2020). While this trend has been reversed during the COVID-19 pandemic, with a 17% reduction in global daily emissions, compared with the average level in 2019, this reduction may be temporary, because no evidence supports any structural changes to the world's industrial systems. Retaliatory emissions are likely to occur after the industrial production activities recover. As such, slowing down global warming by reducing industrial emissions remains a top priority among long-term strategies for global environmental governance (Fang & Heijungs, 2015).

China is the largest emerging economy worldwide, which accounted for approximately 30% of the global carbon emissions in 2019 (Shan et al., 2020). Unlike most of the G20 nations that aim to reduce their national emissions to a certain level in their individual Nationally Determined Contributions (NDCs), China has announced its aim to reach peak emissions in its NDCs by 2030 (NDRC, 2015), and further become carbon neutral by 2060 (Mallapaty, 2020). Officially, China's total emission allowance before 2030 is not yet determined; it has been allocated to different provinces and cities where emission reduction actions are primarily implemented (Fang et al., 2019b). As China has planned to sustain its economic growth through the massive development of infrastructure and investment in the manufacturing sector (The State Council of China, 2020), the likelihood that China's emissions will peak in the coming decade becomes increasingly uncertain. However, China must reach its peak emissions to fulfilling the global 1.5°C or 2°C goals, which has attracted enormous attention from academia over the past few years. The estimates, however, often appear to be divergent, even contradictory

DOI: 10.4324/9781003433859-4

(Fang et al., 2019a; Jiang et al., 2019a). Many argue that China's carbon emissions may peak at the national level by 2030, more likely between 2025 and 2030 (Gallagher et al., 2019; Song et al., 2018), while others object to this argument (Yuan et al., 2014; Zhou et al., 2014).

Given the dominating role of industrial-specific emissions in the national emissions inventory, several studies have focused on the peaking paths of several industrial sectors, such as transportation (Zhou & Zhao, 2016), textile (Huang et al., 2017), and building sectors (Xiao et al., 2014). Further, China has entered an advanced stage of industrialization (Yu et al., 2018); emissions in industrial sectors are expected to peak first, without which a nationwide peaking of emissions would be impossible. This highlights the complexity and diversity of various industrial sectors. For instance, energy-related emissions in energy-intensive industries are more likely to peak faster, compared with those in light industries (Jiang et al., 2019b). Specifically, emissions in the cement and steel sectors are predicted to peak by 2020 (Chen et al., 2018; Gao et al., 2017; Zhang et al., 2018a); those in the electricity sector, between 2020 and 2030 (Viebahn et al., 2015; Yang et al., 2016); those in the petrochemical sector, around 2030 (Lin & Wang, 2019); and those in the transportation and building sectors, between 2030 and 2035 (Pan et al., 2018; Tan et al., 2018). The industrial sectors' diversified emission trajectories increase the complexity and uncertainty of China's goal to have peak emissions before 2030 (Shao et al., 2016). Therefore, narrowing the focus to a single or few sectors is therefore insufficient to represent industrial emissions that substantially vary across sectors (Chen et al., 2020). For this reason, we argue that a multi-sectoral perspective can provide novel insights into the potential peak emission pathways in China. However, to the best of our knowledge, future emissions between various industrial sectors in China have not been analyzed comparatively in the existing literature, and no studies translate China's peaking goal into the sectoral levels.

To predict future emissions and peaking paths, the Environmental Kuznets Curve (EKC) hypothesis can be examined in many ways using econometric models (Al-Mulali et al., 2015). Furthermore, by incorporating scenario analysis into the econometric study, the possible emission trends can be understood (Fang et al., 2019a). Scenario analysis can be broadly divided into two categories: static and dynamic (Lin &Liu, 2010). Current scenario analyses often include arbitrary scenario settings that assume a fixed rate of changes in variables (Jiang et al., 2019b). Despite the broad spectrum of applicability, this approach is flawed in that quantitative uncertainty analysis is either lacking or replaced by sensitivity analysis, which can create an overly simplistic estimation of uncertainty (Wang & Liang, 2013). The omission of uncertainties cannot be ignored; specifically, if some extreme scenarios rarely occur, they should not be treated like other scenarios.

To overcome the shortcomings of static scenario analysis, dynamic scenario analysis was employed. As a mainstream dynamic scenario analysis, Monte Carlo simulation has been envisaged to quantify uncertainty by distributing probabilities to key variables, based on historical data or expert judgments (Vithayasrichareon & MacGill, 2012). Unlike deterministic approaches, Monte Carlo simulation can be perceived broadly as a stochastic method that is appropriate for addressing problems that involve diverse, and even conflicting, uncertainties (Zhang et al., 2017). Monte Carlo simulation has become increasingly popular over the past few years because of uncertainty testimonies (Cooke, 2015), with applications at scales ranging from a single sector (Lorenzo et al., 2012), cities (Wang et al., 2019), and nations (Li & Lin, 2016), among others. (Wang et al., 2019). By providing a scientifically robust methodology for predicting different variables with uncertainty range, Monte Carlo simulation facilitates the identification of likely emission trajectories, instead of the fixed assumptions of some variables (Liu & Lin., 2010; Shao et al., 2017).

In this study, we investigate the emission trajectories of various industrial sectors in China using both regression analysis and Monte Carlo simulation to identify desirable pathways for different sectors to reach peak emissions. This sector-level investigation would also be informative for other nations, such as India—the country that is expected to replace China as the world's largest carbon emitter in the coming years. In summary, the novel contributions of this study to the literature are threefold: (1) they present a full picture of China's future emission trajectories for eight industrial sectors for the first time. (2) This study combines regression analysis and Monte Carlo simulation to examine the EKC hypothesis with transparent and comprehensible uncertainty estimates, and (3) the focus on peaking emissions is shifted from the national level to the sectoral level, which is key in achieving the NDCs' goals that have been neglected for a long time.

4.2 Methods and data

4.2.1 Sectoral carbon emissions accounting

This study accounts for China's carbon emissions that are relevant to energy consumption at both sectoral and national levels using the following equation:

$$C_s = \frac{44}{12}\sum_{i=1}^{4}(F_{s,i}E_{s,i}) \tag{4.1}$$

where C_s is the carbon emissions of sector s, F_i^s is the carbon emission factors for energy I, and $E_{s,i}$ is the amount of consumption of sector s in energy i. Energy consumption could be classified into four types: coal, oil, natural gas, and non-fossil, all of which are then converted into standard coal, as per their respective heat values (Yang et al., 2018). The emission factors are obtained from China's Energy Research Institute (Wu et al., 2018).

4.2.2 The EKC hypothesis and regression analysis

We employ a quadratic regression model to test the existential EKC hypothesis for the carbon emissions and GDP per capita in each sector using the following equation:

$$lnC_s = a + bln\text{GDP}_{st} + c(\text{GDP}_{st})^2 + \varepsilon_s \tag{4.2}$$

where GDP_{st} is the GDP per capita at time t of sector s; a, b, and c are the coefficients that need to be determined; and ε is an error term representing all the other variables that may affect emissions.

While ordinary least square regression is applied to the parameter estimation typically, its ability to deal with the multicollinearity of different independent variables is limited (Jalil & Mahmud, 2009). Furthermore, if time-series data lack co-integration, the regression results of co-integration methods such as the benchmark regression may be biased (Zhang et al., 2018b). The autoregressive distributed lag (ARDL) method has been broadly adopted in many research fields (Baek, 2015; Mrabet & Alsamara, 2017; Zhang et al., 2018a). The advantages of the ARDL method, over other regression methods, include (1) a lower requirement of time-series data for sample size and no requirement for robustness testing; and (2) it can provide unbiased and valid evaluation, even if the independent variables are endogenous (Banerjee et al., 1993; Laurenceson & Chai, 2003). As such, the ARDL method is employed in this study for regression analysis using the following equation:

$$\Delta lnC_{st} = a_0 + \sum_{k=1}^{n1} a_{k1}\Delta lnC_{s,t-k} + \sum_{k=1}^{n2} a_{k2}\Delta ln\mathrm{GDP}_{s,t-k} + \sum_{k=1}^{n3} a_{k3}\Delta(ln\mathrm{GDP}_{s,t-k})^2 + \delta_1 lnC_{s,t-1} + \delta_2 ln\mathrm{GDP}_{s,t-1} + \delta_3(ln\mathrm{GDP}_{s,t-1})^2 + \varepsilon_{st} \tag{4.3}$$

where $C_{s,t}$, $C_{s,t-k}$, and $C_{s,t-1}$ are the carbon emissions of sector s at time t, t-k, and t-1, respectively. $C_{s,t-k}$ is the carbon emissions of sector s at time t-k; $\mathrm{GDP}_{s,t-k}$ and $\mathrm{GDP}_{s,t-1}$ are the GDP per capita of sector s at time t-k and t-1, respectively; δ_1, δ_2, and δ_3 are the coefficients of the lags of $lnC_{s,t-1}$, $ln\mathrm{GDP}_{s,t-1}$, and $(ln\mathrm{GDP}_{s,t-1})^2$, respectively; a_0 and ε_{st} are the constant and error terms, respectively; and a_{k1}, a_{k2}, and a_{k3} are the parameters to be determined.

4.2.3 *Monte Carlo simulation*

Scenario analysis serves as a valuable tool for projecting carbon emissions pathways at multiple scales (Zhang et al., 2017). In this study, Monte Carlo simulation is performed as a scenario analysis that forecasts the carbon emissions from China's eight sectors. The Monte Carlo simulation process consists of three steps. First, we define prior probabilities for critical variables that drive carbon emissions, such as GDP per capita. As the probability distribution of these variables is not precisely known, randomly selecting the rates at which the drivers change is more suitable for applying the triangular distribution function (Ramírez et al., 2008). Second, multiple simulations are conducted by distributing a random sample variable range based on the pre-defined probability. In our case, each industrial sector implements 100,000 simulations, as a more significant number of simulations produces more accurate results. Third, the results of GDP per capita simulation, accompanied with frequency distributions, provide possible output values. By doing so, Monte Carlo simulation facilitates a more accurate evaluation of the future GDP per capita and carbon emissions for each industrial sector in China.

4.2.4 Data

For consistency, we collected the data of China's 56 sectors for the period 1995–2017 and aggregated them into eight major industrial sectors: *Agriculture*; *Building*; *Business*; *Manufacturing*; *Mining*; *production and distributions of electricity, heat, gas, and water (PDEHGW)*; *Building*; *Others;* and *Transportation* as per the new Chinese industrial classification standard in 2002. The components of the eight industrial sectors are displayed in Table 4.1. Data sources are presented in Table 4.2. The industrial value-added data are derived from the China Industry

Table 4.1 Classification of China's eight industrial sectors is composed of 56 sub-sectors

No.	*Industrial sectors*	*Sub-sectors*
1	*Agriculture*	Agriculture, Farming of animals, Farming of animals and fishing, Fishing, Forestry, Forestry, Service activities for agriculture
2	*Building*	Architectural decoration, Architectural installation, Construction of building & Civil engineering, Other construction
3	*Business*	Accommodation and restaurants, Wholesale and retail trade
4	*Manufacturing*	Agricultural non-staple food processing, Beverage production, Chemical fiber products, Cultural, educational and sports articles, Electrical machinery and equipment, Food production, Furniture manufacturing, Papermaking and paper products, Garments and other fiber products, General purpose machinery, Leather, furs, down and related products, Measuring instrument machinery, Medical and pharmaceutical products, Metal products, Nonmetal mineral products, Petroleum and nuclear fuel processing and coking, Printing and recording medium reproduction, Raw chemical materials, and chemical products, rubber and plastic products, Smelting and pressing of ferrous metals, smelting, and pressing of nonferrous metals, Special purpose machinery, Telecommunication, and other electrical equipment, Textile industry, Timber processing, bamboo, cane, palm fiber and straw products, Tobacco processing, Transportation equipment
5	*Mining*	Coal mining and dressing, Ferrous metals mining and dressing, Nonferrous metals mining and dressing, Nonmetal minerals mining and dressing, Petroleum and natural gas extraction
6	*PDEHGW*	Production and distribution of electricity and heat power, Production and distribution of gas, Production, and distribution of water
7	*Others*	Other sectors in the tertiary industry, such as Culture, sports and entertainment, Education, Finance, Information transmission, software, and information technology, Professional technique services, Public management, social security and social organization, Real estate Tenancy and business, Research, and experimental development, Sanitation, social security and social welfare, Scientific research and technical service, Scientific research, technical service and geologic perambulation, Service to households, repair, and other services
8	*Transportation*	Transportation, Storage, and Post

Table 4.2 Data sources for critical variables

Variable	*Description*	*Data sources*
Carbon emissions	Total carbon emissions from energy consumption	China Energy Statistical Yearbooks (1996–2018)
GDP per capita	Total GDP of a sector divided by end-year employment of the sector	China Economic Census Yearbook (2004), China Industrial Statistical Yearbooks (1996–2018) China Industry Economy Statistical Yearbooks (1996–2018) China Statistical Yearbooks (1996–2018)
Population	Employment by sector at the end of a year	China Statistical Yearbooks (1996–2018)

Economy Statistical Yearbooks (1996–2018) and China Economic Census Yearbook (2004). The data on GDP has been converted to compare prices in 2010.

4.3 Results

4.3.1 Carbon emissions accounting

In the past decades, industrialization has driven the rapid increase in China's carbon emissions. Emissions from heavy industrial sectors, such as the *Manufacturing, Mining,* and *PDEHGW* sectors, have risen from 0.21 to 1.81 billion tons from 1995 to 2017 (Figure 4.1), accounting for more than 70% of the total emissions from the eight sectors. Of these, over 56% of the emissions can be attributed to the *Manufacturing* sector, followed by the *PDEHGW*, *Transportation*, and *Mining* sectors, with shares of 30%, 5%, and 5% in 2017, respectively. In contrast to the decreasing trend in the emissions from the *Manufacturing* and *Mining* sectors, the increasing emissions from the *Agriculture, Building, Transportation,* and *Others* sectors are observed. As of 2017, the remaining industrial sectors, namely *Others* (1.8%), *Agriculture* (0.9%), *Business* (0.4%), and *Building* (0.8%), collectively made up approximately 4% of the total values.

For the emissions per capita (Figure 4.2), the *Manufacturing, Mining, PDEHGW*, and *Transportation* sectors are the top four emitters, with an average annual growth rate of 7%, 5%, 14%, and 17%, respectively. The lowest and highest annual growth rates are found in *Agriculture* and *Building* sectors, with an estimated 2% and 18%, respectively. The *Business* and *Other* sectors, often classified as the tertiary industry, have lower emissions per capita, compared with the rest of the sectors. As of 2017, the *Manufacturing* sector was producing 145 times the emissions per capita of the *Business* sector.

4.3.2 The EKC hypothesis testing

The first step in analyzing time-series data is to determine whether they are stationary. To achieve this goal, we chose the generalized least-squares detrended

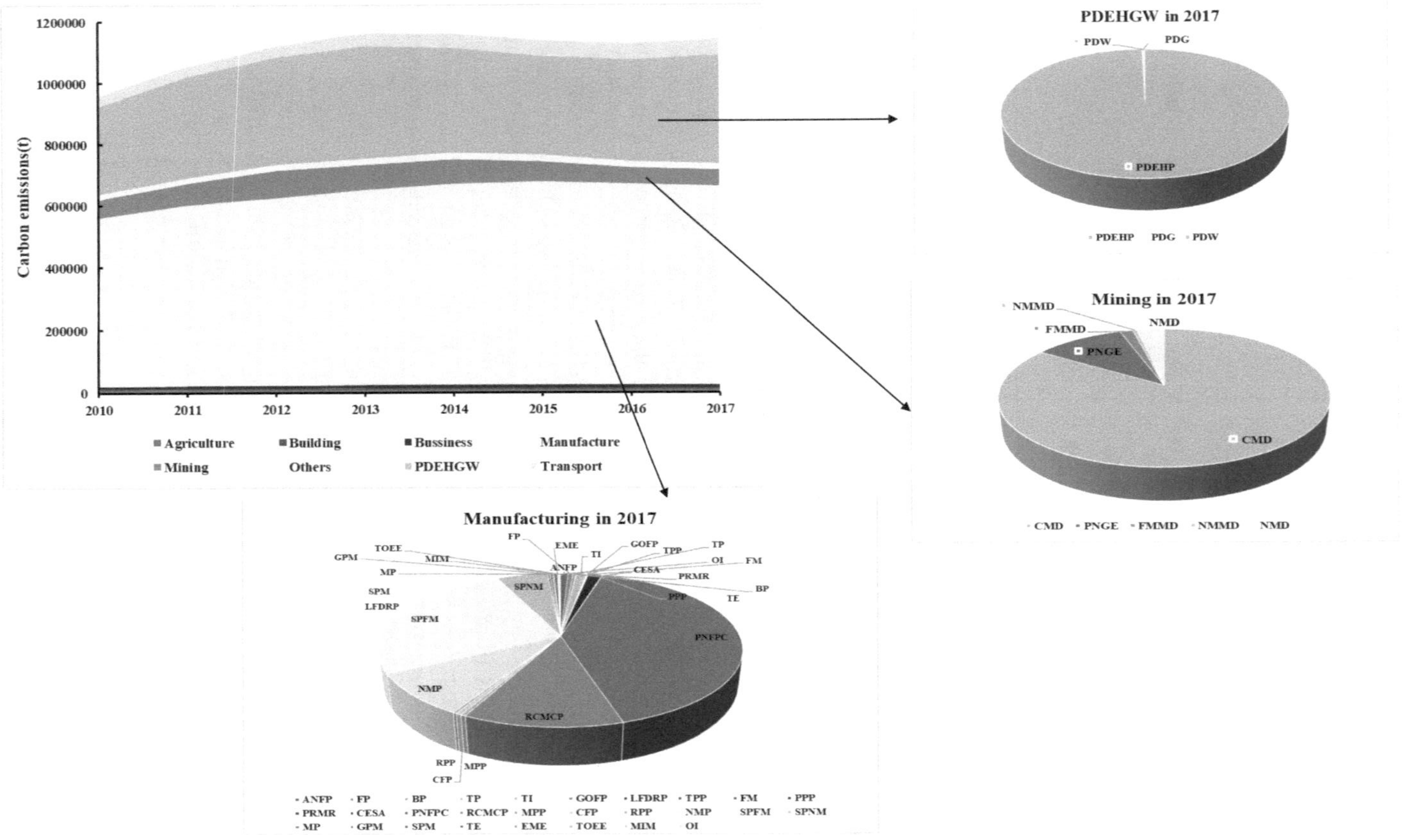

Figure 4.1 Trends in the carbon emissions of eight industrial sectors

Note: We also provide a detailed sub-sectoral composition of the top-three emitters. Abbreviations of these sub-sectors can be found in Appendix B, Table B.1.

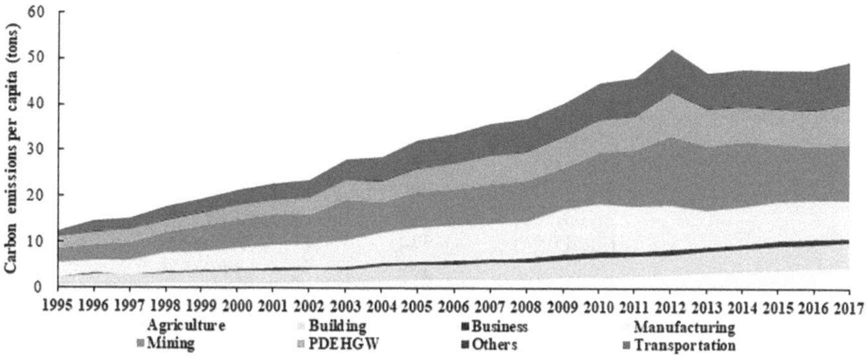

Figure 4.2 Trends in the carbon emissions per capita of eight industrial sectors

augmented Dickey–Fuller test (DF-GLS) unit root tests because it is more suitable for small samples, compared to other unit root tests. The results displayed in Table 4.3 show that the logarithmic sequence of three variables (i.e., carbon emissions per capita, GDP per capita, and the square of GDP per capita) for all eight sectors are stable at the 1% or 5% significance levels. In contrast, the logarithmic sequence of the variables for the *Building, Business, Manufacturing, Mining, PDEHGW*, and *Transportation* sectors are stable at the 5% or 10% significance levels. In sum, all the variables used in this study are consistent with I (0) or I (1) stationery. Therefore, they can be used for the estimation of the ARDL model.

The ARDL bound test is appropriate for small-sized samples with practical estimation and high tolerance of bound (Pesaran & Shin, 1995; Pesaran et al., 2001). The ARDL bound test is used to examine the co-integrating relationships among the eight sectors. First, it is necessary to determine the optimal lag length based on the Akaike information criterion (AIC). Table 4.4 shows the results of lag order selection.

We test the EKC hypothesis, in which the carbon emissions per capita of China's eight industrial sectors are dependent variables. After introducing the panel data for the period 1995–2017 into the ARDL method, the corresponding sectoral GDP per capita and its square become independent variables. The regression results for all the sectors are displayed in Table 4.5. As shown, the EKC hypothesis holds for seven industrial sectors, in which GDP per capita exhibits an inverted U-shaped relationship with sectoral emissions per capita. We find that the GDP per capita has the most significant impact on the emissions from the *Agriculture* sector, followed by the *Building* and *Transportation* sectors. The emissions from the *Manufacturing*, *Mining*, *PDEHGW*, and *Others* sectors correlate significantly with economic growth. Regarding the *Business* sector, however, no evidence supports the EKC hypothesis. Instead, a linear correlation has been observed, implying that the *Business* sector is unlikely to reach peak emissions in the coming years. This suggests that there is much room for the development of the *Business* sector, partially because most of the sub-sectors, such as the Accommodation and restaurants and Wholesale and retail trade in the *Business* sector, appear to be environmentally

Table 4.3 Unit root test (DF-GLS) for eight industrial sectors

Sectors	*Test var. (x)*	*Level (T-statistic)*	*1st diff. (T-statistic)*
Agriculture	lnC	−0.048	−3.264***
	lnGDP	−0.163	−2.643**
	$(\ln GDP)^2$	−0.090	−2.724***
Building	lnC	−1.739*	−4.208***
	lnGDP	−2.385	−3.612***
	$(\ln GDP)^2$	−1.534	−3.208***
Business	lnC	−2.592	−4.808***
	lnGDP	−3.761***	−2.048**
	$(\ln GDP)^2$	−2.958***	−3.724***
Manufacturing	lnC	−2.307**	−2.753**
	lnGDP	−1.417	−2.241***
	$(\ln GDP)^2$	−2.768**	−2.038***
Mining	lnC	−2.095**	−4.828***
	lnGDP	−1.110	−2.092***
	$(\ln GDP)^2$	−0.182	−4.255***
PDEHGW	lnC	−0.827	−4.211***
	lnGDP	−0.750	−4.088***
	$(\ln GDP)^2$	−1.792**	−4.238***
Others	lnC	−0.744	−5.039***
	lnGDP	−0.735	−3.705***
	$(\ln GDP)^2$	−0.067	−3.608***
Transportation	lnC	−0.827	−4.211***
	lnGDP	−2.798**	−4.062***
	$(\ln GDP)^2$	−1.792*	−4.238***

Note: ***, **, and * represent statistical significance at the 1%, 5%, and 10% level, respectively.

Table 4.4 ARDL bound test co-integration for eight industrial sectors

	Sectors	*Optimal AIC lags*	*F-statistics*	*Co-integration exist or not*
Estimated model FC (LLC/LnGDP, $LnGDP^2$)	*Agriculture*	(2,1,1)	4.551**	Yes
	Building	(2,1,2)	22.891***	Yes
	Business	(2,1,0)	5.135**	Yes
	Manufacturing	(1,1,0)	17.792***	Yes
	Mining	(1,2,1)	12.436***	Yes
	PDEHGW	(1,1,0)	14.516***	Yes
	Others	(1,1,0)	35.197***	Yes
	Transportation	(1,1,2)	8.126***	Yes

	Critical values of Pesaran et al. (2001), N = 30	
Significance level	*Lower bound I (0)*	*Upper bound I (1)*
1%	4.614	5.966
5%	3.272	4.306
10%	2.676	3.586

Notes: *** and ** represent statistical significance at the 1% and 5% level, respectively. The critical values are collected from Pesaran et al. (2001). AIC determines the optimal lag length.

Table 4.5 ARDL regression analysis for eight industrial sectors

Sectors	*LnGDP*	*(LnGDP)²*	*EKC holds*	*Turning point (RMB)*	*GDP per capita in 2017*
Agriculture	3.573*** (0.891)	−0.170*** (0.048)	Yes	37,281	29,650
Building	6.555** (3.293)	−0.308** (0.172)	Yes	46,631	20,927
Business	3.803* (2.964)	0.862* (0.542)	No	None	13,249
Manufacturing	0.381* (0.094)	−0.017* (0.621)	Yes	71,396	51,883
Mining	21.829* (16.104)	−1.012* (0.778)	Yes	55,050	45,177
PDEHGW	2.408* (2.263)	−0.107* (0.058)	Yes	80,821	57,941
Others	1.424** (0.473)	−0.064** (0.146)	Yes	86,452	39,991
Transportation	2.371*** (0.738)	−0.088** (0.039)	Yes	53,236	44,048

Notes: ***, **, and * represent statistical significance at the 1%, 5%, and 10% level, respectively. Standard errors are provided in parentheses.

friendly—they do not involve excessive material production and burning of fossil fuels (Hou et al., 2020).

In contrast to the *Business* sector, the emissions in all the other industrial sectors tend to peak when the corresponding GDP per capita reaches a turning point. The *Agriculture* sector was observed to have the lowest turning point, with a GDP per capita of 37,281 RMB. This is followed by the *Building*, *Transportation*, *Mining*, *Manufacturing*, and *PDEHGW* sectors. The *Others* sector had the highest turning point, with a GDP per capita of 86,452 RMB. Furthermore, we estimate the number of years that are required for each sector to reach the peak emissions at an average growth rate over the past 13 years. The gaps between current GDP per capita and estimated turning points vary substantially across industrial sectors, ranging from 7,631 RMB for the *Agriculture* sector to 46,461 RMB for the *Others* sector in absolute values. Such variations warrant an in-depth discussion on the differentiated pathways to peak China's emissions at the sectoral level, rather than narrowing the focus to the whole country or considering various industries as one sector.

4.3.3 Monte Carlo simulation

Considering all the economic growth scenarios under the existing policies, we explore the probability distributions of GDP per capita through Monte Carlo simulation over 100,000 times. By exploring the correlation of peak emissions per capita and the corresponding GDP per capita further, we noted prominent differences between different industrial sectors. As mentioned above, the highest turning point (86,452 RMB) is observed in the *Others* sector, and it is 2.3 times the lowest turning point (37,281 RMB) for the *Agriculture* sector. However, this does not necessarily mean that reaching peak emissions in the *Others* sector is more complex than it is in the other cases, because the sectoral GDP per capita and its growth rates vary considerably. In 2017, for instance, the current GDP per capita of the *Agriculture* and *Others* sectors amounted to 29,650 and 39,991 RMB, accounting for 80% and 46% of the turning points, respectively. By 2030, the GDP per capita is most likely to rise to 45,972 and 64,268 RMB in the *Agriculture* and *Others* sectors, respectively. The simulation results indicate that, in all the industrial sectors, the GDP per capita will increase continuously, although the growth will decrease. This is particularly true for the *Mining* and *PDEHGW* sectors, in which the GDP per capita will increase with average annual rates of 8% and 6% between 2018 and 2024, and 6% and 4% between 2024 and 2030, respectively.

4.3.4 Carbon emissions projection

China's future per capita emission trajectories for all the industrial sectors, except the *Business* sector, are illustrated in Figure 4.3. The results show a significant variation in the peak-reaching time and emissions among these sectors. The *Building* and *Others* sectors are expected to reach their peak emissions per capita in the

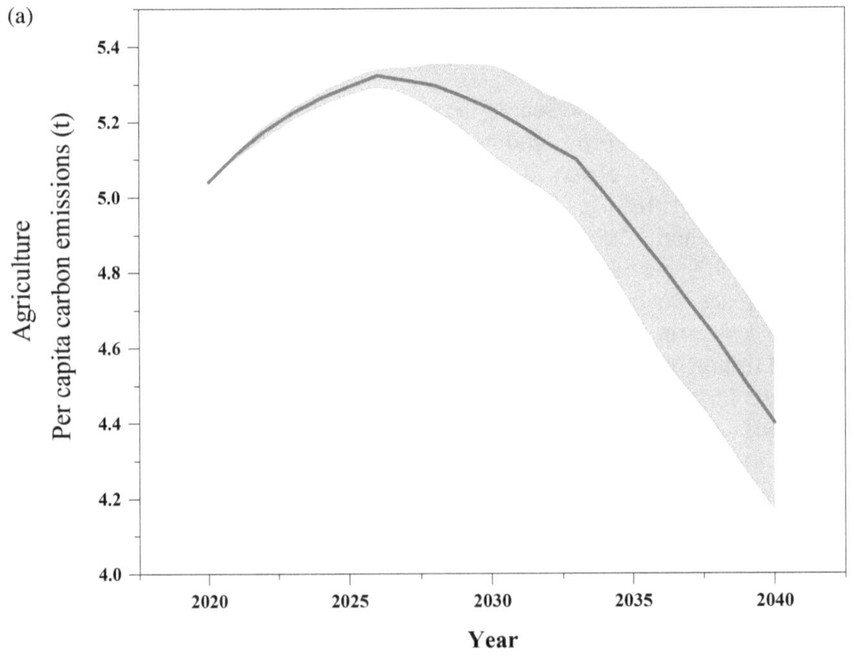

Figure 4.3 Average values and 95% confidence intervals of the prediction of sectoral emissions per capita through Monte Carlo simulation (a) *Agriculture* (b) *Building* (c) *Manufacturing* (d) *Mining* (e) *PDEHGW* (f) *Others* (g) *Transportation* (*Continued*)

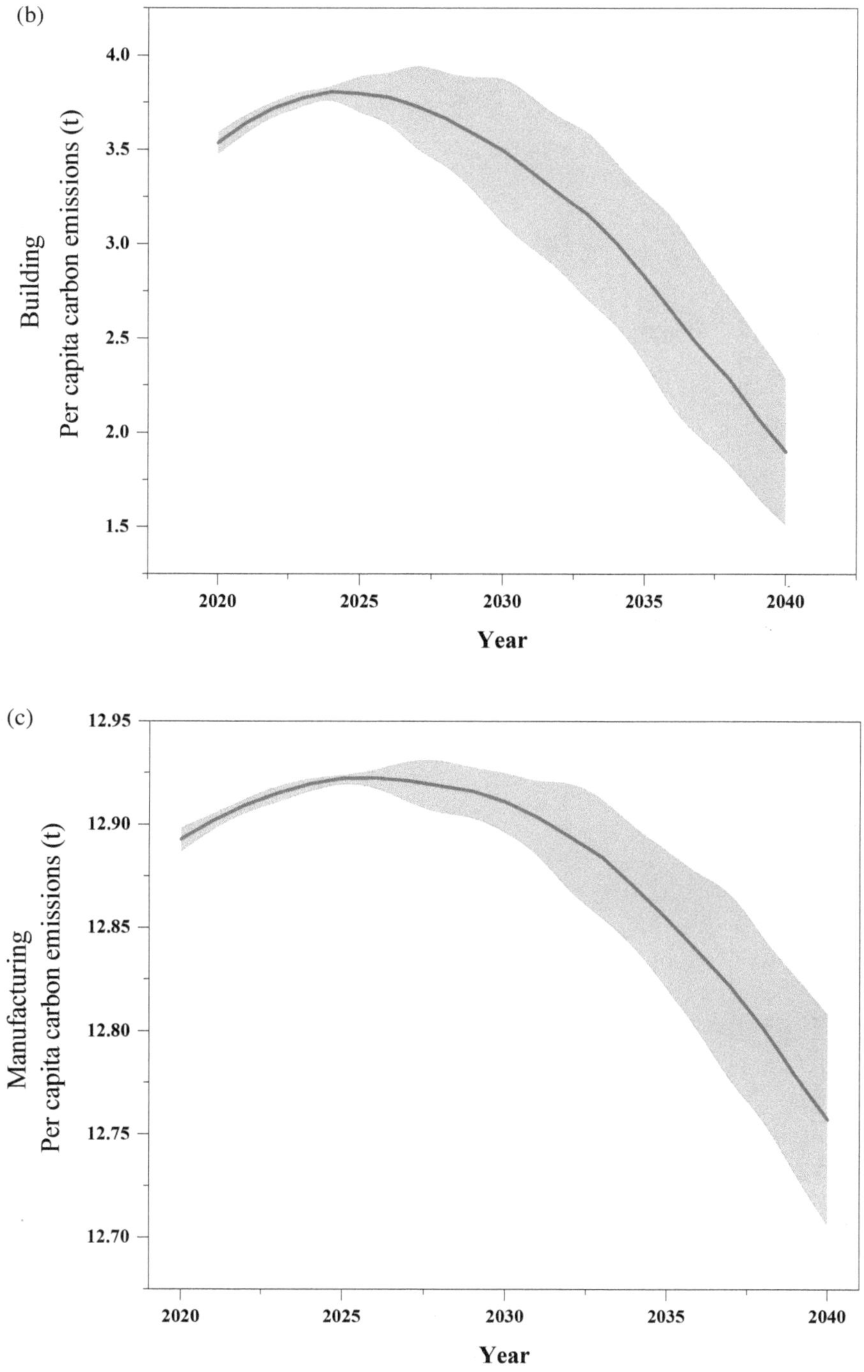

Figure 4.3 (Continued)

(d)

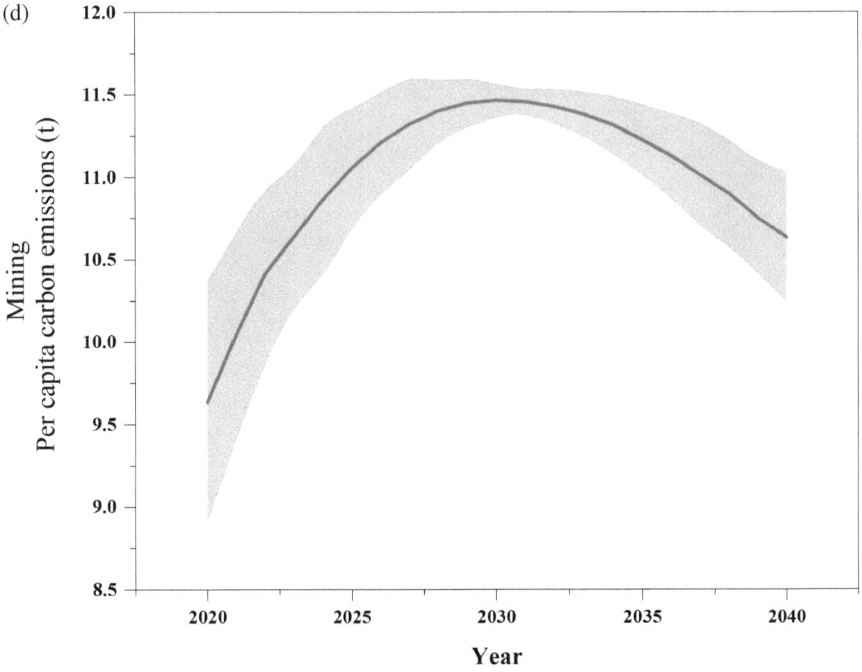

(e)

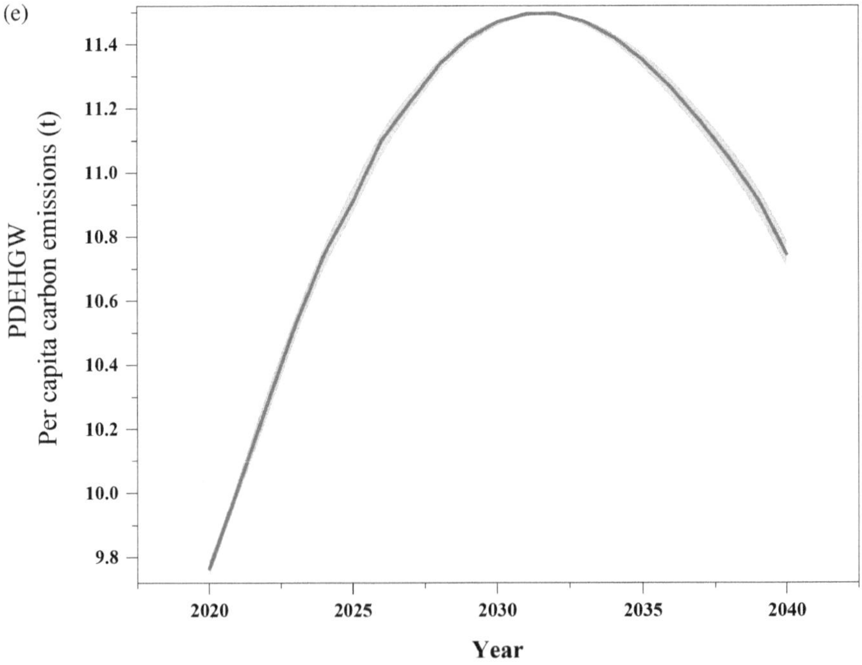

Figure 4.3 (Continued)

(f)

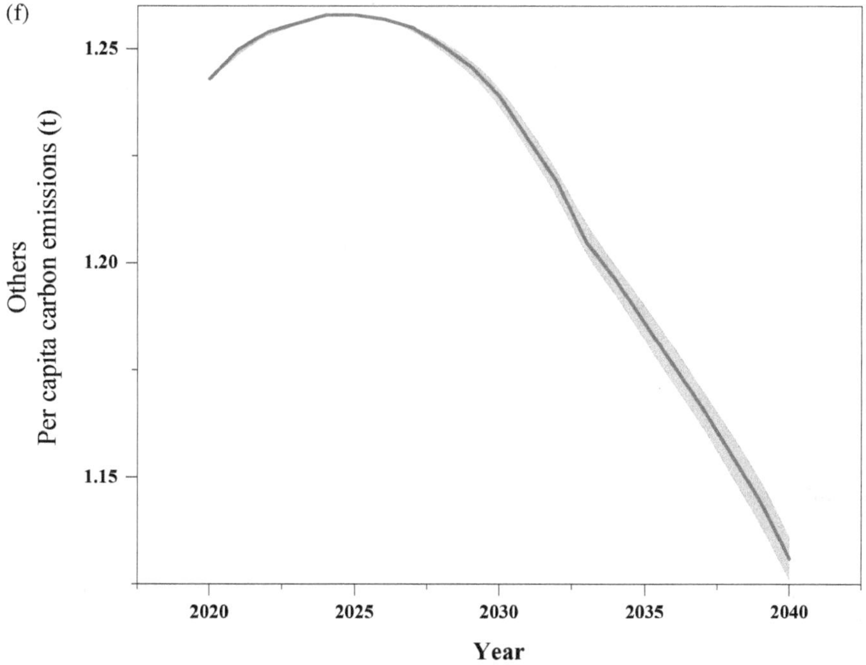

(g)

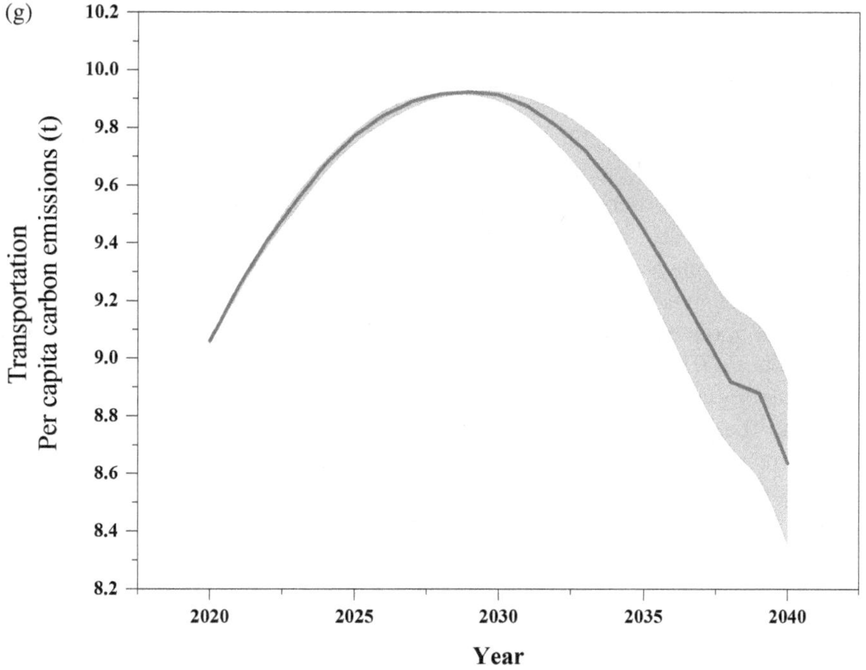

Figure 4.3 (Continued)

period between 2021 and 2028, with the possible peak values of 3.7 and 1.25 tons, respectively. However, as GDP per capita continues to grow, the *Building* sector will experience a much more rapid decline than the *Others* sector after reaching their peak emissions. Carbon emissions per capita from the *Agriculture, Transportation*, *Manufacturing*, *PDEHGW,* and *Mining* sectors are expected to peak during the periods 2025–2032, 2025–2034, 2026–2036, 2028–2033, and 2029–2034, respectively. Nevertheless, the discrepancy in the emission trajectories and peak values between these industrial sectors seems remarkably large. The *Manufacturing*, *PDEHGW,* and *Transportation* sectors would maintain a relatively stable emission level after reaching the peak. In contrast, a relatively sharp decrease in the emissions per capita can be observed for the *Agriculture* and *Mining* sectors. Among the seven industries, the highest peak emissions per capita are found in the *Manufacturing* and *Mining* sectors at a range of 12.76–12.88 and 9.69–11.46 tons, respectively. They are followed by the *PDEHGW* and *Transportation* sectors, whose peak emissions per capita are estimated at 9.74–13.42 and 8.71–9.86 tons, respectively. The anticipated peak emissions per capita of the *Agriculture*, *Building*, and *Others* sectors appear to be much lower, ranging from 5.08 to 5.44, 1.46 to 3.71, and 1.13 to 1.26 tons, respectively.

Manufacturing, *PDEHGW*, and *Others* sectors are under more significant pressure, and they need to reach the peak emissions when the GDP per capita amounts to 71,396, 80,821, and 86,452 RMB, respectively. *PDEHGW* has a lower peak, which means that this industry should further promote low-carbon production while maintaining good economic growth. The peak GDP per capita of the *Agriculture* and *Construction* sectors is relatively low, both below 50,000 RMB. However, due to these two industries' low peak emissions per capita, the pressure to reach the peak is also more significant. The GDP per capita of the *Transportation* and *Mining* sectors reached 53,236 and 55,050 RMB, respectively, when the inflection point was reached, and the peak emissions per capita of the *Transportation* sector were relatively high; therefore, the pressure of the *Transportation* sector to reach the peak was relatively small in comparison.

Figure 4.4 displays the earliest, latest, and most likely peaking time and values of seven sectors in total. The *Agriculture* sector is likely to peak before 2030, most likely in 2025, owing to the ever-improving technology in agricultural production (Zhang et al., 2019). The forecast results indicate that the *Building* sectors' GDP will increase rapidly between 2018 and 2023 and slowly decline between 2023 and 2025, with the peaking period ranging from 2021 to 2028, most likely in 2024, which corresponds to 3.70 tons. Meanwhile, the *Manufacturing* and *Mining* sectors will peak in the periods 2026–2036 and 2029–2034, with peaking values of 12.89 tons and 11.46 tons, respectively. Because the *Manufacturing* sector accounts for 62% of total emissions, this sector plays a crucial role in the expected peaking of emissions among all the industrial sectors (Wu et al., 2019a). Both the *Manufacturing* and *Mining* sectors have significant probabilities of peak emissions per capita before 2030. If the above two sectors could achieve a substantial reduction in emissions, China will likely become emissions free sooner than previously thought. If not, China is more likely to undergo a long-term plateau after its carbon emissions

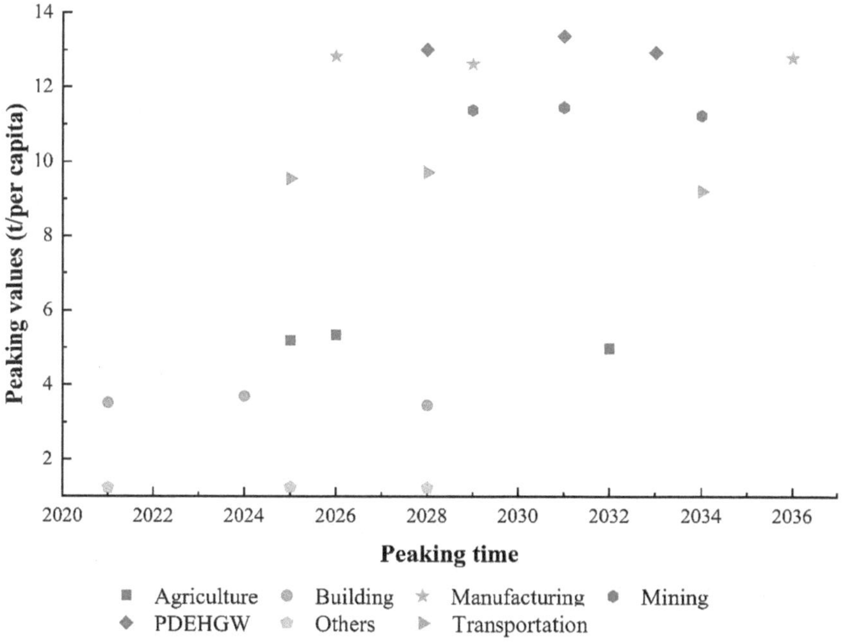

Figure 4.4 Peaking time of seven industrial sectors with three scenarios

peak. The *Others* sector will peak between 2021 and 2028, most likely in 2025, with a peaking value of 1.25 tons. However, emissions from the *PDEHGW* sector will peak between 2028 and 2033, most likely in 2031, with a peaking value of 13.38 tons. The *Transportation* sector is considered as one of the fastest increasing drivers in China (Li et al., 2019) and is predicted to peak between 2025 and 2034, most likely in 2028, with a peaking value of 9.72 tones.

4.4 Discussion

4.4.1 Comparison with literature

Previous studies have increasingly focused on the pathway projections of China's peak emissions. A comparison between the peaking periods of industrial sectors in China in this study and those in previous studies is provided in Table 4.6. As illustrated, the various sectors differ dramatically in their possible peak-reaching time.

Our prediction estimates on the *Agriculture*, *Manufacturing*, *PDEHGW*, and *Mining* sectors are generally consistent with those in the existing literature. Meanwhile, we find that it is easier for the emissions in the *Building* and *Transportation* sectors to peak, contrasting previous predictions. The most notable gaps between our estimates and those in the literature are observed in the *Transport* and *Building* sectors. Owing to the exponentially growing use of electric vehicles and

Table 4.6 Comparison of this paper with the previously researches

Sector	*Reference*	*Peaking year with the highest probability*	*The earliest peaking year*	*The latest peaking year*
Agriculture	This paper	2026	2025	2032
	Chen et al., 2020		2021	2031
Building	This paper	2024	2021	2028
	Ma et al., 2019	2031	2027	2036
Manufacturing	This paper	2029	2026	2036
	Shao et al., 2017	2031	2024	2035
Mining	This paper	2031	2029	2034
	Jiang et al., 2019a	2024	2022	2029
PDEHGW	This paper	2031	2028	2033
	Yuan et al., 2018		2023	2031
Transportation	This paper	2028	2025	2034
	Gambhir et al., 2015	2032	2026	2033

low-carbon building materials in recent years, these two sectors will reach peak emissions earlier than previously thought (Gambhir et al., 2015; Ma et al., 2019).

Methodological differences in emission projections are probably the leading cause of the distinction between our results and those in the literature. Generally, the methods used could be divided into two types: econometrics models (e.g., EKC and Kaya identity) and the partial or general equilibrium models (i.e., computable general equilibrium (CGE) and an integrated assessment model (IAM)). Both models have strengths and weaknesses. The projections of the industrial sectors' carbon emissions face a major threat due to the high complexity and randomicity in the economic and social development (Guan et al., 2018). Scenario settings in the existing studies usually involve a certain probability that cannot account for the stochastic uncertainties. Consequently, we employ a sophisticated technique—Monte Carlo simulation—to address the stochastic uncertainties in our emissions prediction.

4.4.2 Innovation and uncertainties

Compared to previous studies, this study makes three novel contributions to the literature. First, we present, for the first time, the full picture of the peaking pathways of the emissions in China's eight industrial sectors, which in aggregate made up 73% of China's total emissions in 2017. Second, in contrast to most scenario analyses that cannot account for stochastic uncertainties, we combine regression analysis and Monte Carlo simulation to examine the EKC hypothesis using probabilistic uncertainty modeling. As a result, the likely emission trajectories are predicted using probability distribution instead of fixed numbers. Third, our paper calls for a shift in the focus from China's peak emissions to the sectoral peak emissions. It proceeds with the recognition that sectors are key in making the efforts to reach peak emissions, which is largely neglected in current discourses on the pathways to peak emissions. Most studies focus on the role of heterogeneity between different regions in reaching the goal of peaking emissions by 2030, while ignoring that between different sectors (He et al., 2018).

Research on future peaking pathways inevitably contains uncertainties and limitations. This study attempts to conduct a preliminary survey of China's emission peaks. First, our accounting of China's carbon emissions by sector has some uncertainties (Gregg et al., 2008; Zhang and Fang., 2019); however, many studies have adopted this approach broadly (Gregg et al., 2008; Guan et al., 2012; Zhang & Chen, 2014). After comparing our estimated emissions with those by Shan et al. (2021), a disagreement of less than 8% is found. Another uncertainty emanates from the variable distribution settings in our Monte Carlo simulation (Green & Stern, 2017). Studies that discuss the possible trends in these variables after the COVID-19 pandemic are scarce. Therefore, future work needs to explore how the energy and economic systems would change in the post-COVID-19 era, both on the international and intranational scales. Moreover, although Monte Carlo has so many advantages in dealing with uncertainty, in this case it does not consider the impact caused by policy change. Therefore, combining specified policy scenarios with Monte Carlo simulations would be a next step toward a more scientifically reliable analysis.

4.4.3 Policy implications

The COVID-19 pandemic has uncovered the fragile economic, social, and environmental underpinnings in the world. China's unprecedented lockdowns against the COVID-19 outbreak have led to a sharp decline in carbon emissions during the early months of the pandemic (Ghosh et al., 2020; Myllyvirta, 2020). While emissions reduction is desirable, this was not the ideal means to achieve this reduction; moreover, the efforts to recover the economy may cause carbon emissions to rebound quickly in the near future (Shan et al., 2021). Although current industrial lobbies tend to pressure the government to relax environmental regulations (Reuters, 2020), climate actions should not take a back burner to economic recovery. The authorities should keenly observe the progress of China's NDCs and propose new industrial, financial, technological, and institutional measures (Ye et al., 2021). Our results have the potential to inform policymakers on sector-specific policies. The *Agriculture* sector is very likely to attain peak emissions before 2030. The challenge entails preventing peak emissions per capita from being too high, and this would require a transition toward a green and modern agricultural system. The same situation applies to the *Others* sector, which is expected to peak emissions prior to 2030 with a relatively low peak-reaching value. The *Transportation* and *Building* sectors are more likely to peak emissions around 2035. The high-quality peaking of emissions in the *Others* sector can be achieved through technological innovation and application; for instance, the development of new energy vehicles (Ye et al., 2021). Regarding the *Building* sector, the share of emissions has increased, and the energy saving techniques for residential buildings should be developed (Wu et al., 2019b). The emissions in the *Manufacturing* sector, which are the highest among the eight sectors, are likely to peak between 2026 and 2036. The more emission reduction policies are implemented, the more likely the peak-reaching time will move forward. The same holds true for the *PDEHGW* and *Mining* sectors, which are supposed to reach peak emissions during the periods 2028–2033 and 2029–2034, respectively. Moreover, energy transition plays a critical role in the ongoing supply-side reform of China. A high proportion of renewables in power systems will enable an earlier peak-reaching time for these sectors (Gielen et al., 2019). Furthermore, developing countries that show continuous increases in carbon emissions play an increasingly important role in global climate mitigation (Debnath & Mourshed, 2018). In that sense, our analysis, which focuses on China—the world's largest emissions trading market and emitter of carbon emissions (Fang et al., 2021)—can serve as a reference for other countries to improve their peaking pathways from a sectoral perspective.

4.5 Conclusions

By tracing China's emission trajectories at the sectoral level and examining the EKC hypothesis for the eight industrial sectors, using both regression analysis and Monte Carlo simulation, we found that: (1) the *Manufacturing*, *Production and distributions of electricity, heat, gas, and water Transportation,* and *Mining* sectors accounted for nearly 56%, 30%, 5%, and 5% of the total emissions from the eight sectors in 2017, whereas the other four sectors collectively contributed 4%. Regarding the emissions per capita, the first two sectors remain unchanged; (2) the *Building, Production and distributions of electricity, heat, gas, and water Transportation,* and *Transportation* sectors experience fast growth in their emissions, while the proportions of the *Manufacturing* and *Mining* sectors continuously decline; (3) the *Agriculture, Building, Transportation,* and *Others* sectors may peak emissions before 2030—the *Manufacturing* and *Production and distributions of electricity, heat, gas, and water Transportation* sectors will more likely peak after 2030, and the peaking pathway of the *Business* sector remains unclear. (4) The differentiated mitigation policies based on the characteristics and dynamics of each industrial sector would be highly needed to fulfill China's latest ambitious goals to peak emissions by 2030 and even become carbon neutral by 2060; however, this analysis field is neglected in current discourses on China's peak emissions.

Future work needs to explore how the energy and economic systems would change in the post-COVID-19 era, both on the international and intranational scales. Although this paper provides novel insights into the potential pathways in which emissions peak in China by presenting a complete picture of the various sectors, rather than single ones, the peak emissions within each of the sectors should be investigated in detail. This is particularly true if the analysis aims to inform policymakers on how to maintain global warming thresholds and improve China's emission trading scheme. For example, sub-sectors like electricity, cement, and chemicals substantially contribute to the national emission inventory; therefore, their individual peaking pathways deserve priority attention. Currently, China's cap and trade is primarily implemented at the provincial level; therefore, studies that focus on the peaking emissions of each sector in different provinces are required

(Zhang et al., 2022). Given the fundamental role of industrial sectors in achieving the Chinese Nationally Determined Contributions, it is necessary to ensure that the long-term mitigation goals of the eight sectors align with China's latest ambition to be carbon neutral by 2060. Research into China's potential peaking pathways at the sectoral level can be of global significance, as it will serve as a reference for other countries facing similar difficulties in identifying the appropriate ways of peaking sectoral emissions.

References

Al-Mulali, U., Saboori, B., & Ozturk, L. (2015). Investigating the environmental Kuznets curve hypothesis in Vietnam. *Energy Policy*, *76*, 123–131.

Baek, J. (2015). Environmental Kuznets curve for CO_2 emissions: The case of arctic countries. *Energy Economics*, *50*, 13–17.

Banerjee, A., Dolado, J. J., Galbraith, J. W., & Hendry, D. (1993). *Co-integration, error correction, and the econometric analysis of non-stationary data*. Oxford University Press.

Chen, Q., Gu, Y., Tang, Z., Wei, W., & Sun, Y. (2018). Assessment of low-carbon iron and steel production with CO_2 recycling and utilization technologies: A case study in China. *Applied Energy*, *220*, 192–207.

Chen, X., Shuai, C., Wu, Y., & Zhang, Y. (2020). Analysis on the carbon emission peaks of China's industrial, building, transport, and agricultural sectors. *Science of The Total Environment*, *709*, 135768.

Cooke, R. M. (2015). Messaging climate change uncertainty. *Nature Climate Change*, *5*, 8–10.

Debnath, K. B., & Mourshed, M. (2018). Challenges and gaps for energy planning models in the developing-world context. *Nature Energy*, *3*, 172.

Fang, K., & Heijungs, R. (2015). The role of impact characterization in carbon footprinting. *Ecology and the Environment*, *13*(3), 130–131.

Fang, K., Tang, Y., Zhang, Q., Song, J., Wen, Q., & Sun, H. (2019a). Will China peak its energy-related carbon emissions by 2030? Lessons from 30 Chinese provinces. *Applied Energy*, *255*, 113852.

Fang, K., Zhang, Q., Song, Y., Yoshida, Y., Sun, L., & Zhang, H. (2019b). How can China achieve its intended nationally determined contributions by 2030? A multi-criteria allocation of China's carbon emission allowance. *Applied Energy*, *241*, 380–389.

Fang, K., Zhang, Q., Song, J., Yu, C., Zhang, H., & Liu, H. (2021). How can national ETS affect carbon emissions and abatement costs? Evidence from the dual goals proposed by China's NDCs. *Resources, Conservation and Recycling*, *171*, 105638.

Gallagher, K. S., Zhang, F., Orvis, R., Rissman, J., & Liu, Q. (2019). Assessing the policy gaps for achieving China's climate targets in the Paris Agreement. *Nature Communications*, *10*, 1256.

Gambhir, A., Tse, L. K. C., Tong, D., & Martinez-Botas, R. (2015). Reducing China's road transport sector CO_2 emissions to 2050: Technologies, costs and decomposition analysis. *Apply Energy*, *157*, 905–917.

Gao, T., Shen, L., Shen, M., Liu, L., Chen, F., & Gao, L. (2017). Evolution and projection of CO_2 emissions for China's cement industry from 1980 to 2020. *Renewable and Sustainable Energy Reviews*, *74*, 522–537.

Ghosh, R., Dubry, M. J., Chatterjee, S., & Dubey, S. (2020). Impact of COVID-19 on children: Special focus on the psychosocial aspect. *Minerva Medica*, *72*(3), 226–235.

Gielen, D., Boshell, F., Saygin, D., Bazilian, M. D., Wagner, N., & Gorini, R. (2019). The role of renewable energy in the global energy transformation. *Energy Strategy Reviews*, *24*, 38–50.
Green, F., & Stern, N. (2017). China's changing economy: Implications for its carbon dioxide emissions. *Climate Policy*, *17*(4), 423–442.
Gregg, J. S., Andres, R. J., & Marland, G.. (2008). China: Emissions pattern of the world leader in CO_2 emissions from fossil fuel consumption and cement production. *Geophysical Research Letters*, 35(8), 1–5.
Guan, D., Liu, Z., Geng, Y., Lindner, S., & Hubacek, K. (2012). The gigatonne gap in China's carbon dioxide inventories. *Nature Climate Change*, *2*(9), 672–675.
Guan, D., Meng, J., Reiner, D. M., Zhang, N., Shan, Y., Mi, Z, Shao, S., Liu, Z., Zhang, Q., & Davis, S. (2018). Structural decline in China's CO_2 emissions through transitions in industry and energy systems. *Nature Geoscience*, *11*, 551–555.
He, W., Yang, Y., Wang, Z., & Zhu, J. (2018). Estimation and allocation of cost savings from collaborative CO_2 abatement in China. *Energy Economics*, *72*, 62–74.
Hou, D., O'Connor, D., Igalavithana, A. D., Alessi, D. S., Luo, J., Tsang, D. C, Sparks, D.L., Yamuchi, Y., Rinklebe, J., & OK, Y. S. (2020). Metal contamination and bioremediation of agricultural soils for food safety and sustainability. *Nature Reviews Earth & Environment*, *1*(7), 366–381.
Huang, B., Zhao, J., Geng, Y., Tian, Y., & Jiang, P. (2017). Energy-related GHG emissions of the textile industry in China. *Resources, Conservation and Recycling*, *119*, 69–77.
IEA. (2020). Global CO_2 emissions in 2019. https://www.iea.org/articles/global-co2-emissions-in-2019
IPCC (2018). *Special report: Global warming of 1.5°C*. Cambridge University Press.
Jalil, A., & Mahmud, S. F. (2009). Environment Kuznets curve for CO_2 emissions: A cointegration analysis for China. *Energy Policy*, *37*(12), 5167–5172.
Jiang, J., Ye, B., & Liu, J. (2019a). Peak of CO_2 emissions in various sectors and provinces of China: Recent progress and avenues for further research. *Renewable and Sustainable Energy Reviews*, *112*, 813–833.
Jiang, J., Ye, B., & Liu, J. (2019b). Research on the peak of CO_2 emissions in the developing world: Current progress and future prospect. *Applied Energy*, *235*(1), 186–203.
Laurenceson, J., & Chai, J. (2003). *Financial reform and economic development in China*. Edward Elgar Publishing.
Li, F., Cai, B., Ye, Z., Wang, Z., Zhang, W., Zhou, P., & Chen, J. (2019). Changing patterns and determinants of transportation carbon emissions in Chinese cities. *Energy*, *174*, 562–575.
Li, K., & Lin, B. (2016). China's strategy for carbon intensity mitigation pledge for 2020: Evidence from a threshold cointegration model combined with Monte-Carlo simulation methods. *Journal of Cleaner Production*, *118*, 37–47.
Lin, B., & Liu, X. (2010). China's carbon dioxide emissions under the urbanization process: Influence factors and abatement policies. *Economic Research Journal*, *45*(08), 66–78.
Lin, B., & Wang, M. (2019). Dynamic analysis of carbon dioxide emissions in China's petroleum refining and coking industry. *Science of the Total Environment*, *671*, 937–947.
Liu, B., & Lin, Y. (2010). China's carbon dioxide emissions under the urbanization process: Influence factors and abatement policies. *Economic Research Journal*, 45(8), 66–78.
Lorenzo, G. D., Pilidis, P., Witton, J., & Probert, D. (2012). Monte-Carlo simulation of investment integrity and value for power-plants with carbon-capture. *Applied Energy*, *98*, 467–478.
Ma, X., Wang, C., Dong, B., Gu, G., Chen, R., Li, Y., Zou, H., Zhang, W., & Li, Q. (2019). Carbon emissions from energy consumption in China: Its measurement and driving factors. *Science of the Total Environment*, *648*, 1411–1420.

Mallapaty, S. (2020). How China could be carbon neutral by mid-century. *Nature*, *586*(7830), 482–483.
McGlade, C., & Ekins, P. (2015). The geographical distribution of fossil fuels unused when limiting global warming to 2°C. *Nature*, *517*, 187–190.
Mora, C., Rollins, R. L., Taladay, K., Kantar, M. B., Chock, M. K., Shimada, M., & Franklin, E. C. (2018). Bitcoin emissions alone could push global warming above 2°C. *Nature Climate Change*, *8*, 931–933.
Mrabet, Z., & Alsamara, M. (2017). Testing the Kuznets curve hypothesis for Qatar: A comparison between carbon dioxide and ecological footprint. *Renewable and Sustainable Energy Reviews*, *70*, 1366–1375.
Myllyvirta, L. (2020). Coronavirus has temporarily reduced China's CO_2 emissions by a quarter. Carbon Brief. Retrieved April 20, 2020, from https://www.carbonbrief.org/analysis-coronavirus-has-temporarily-reduced-chinas-co2-emissions-by-a-quarter.
NDRC (2015). *The People's Republic of China first biennial update report on climate change of China*. China Planning Press.
Pan, X., Wang, H., Wang, L., & Chen, W. (2018). Decarbonization of China's transportation sector: In light of national mitigation toward the Paris Agreement goals. *Energy*, *155*, 853–864.
Pesaran, M. H., & Shin, Y. (1995). *An autoregressive distributed lag modeling approach to co-integration analysis*. RePEc.
Pesaran, M. H., Shin, Y., & Smith, R. J. (2001). Bounds testing approaches to the analysis of level relationships. *Journal of Applied Econometrics*, *16*(3), 289–326.
Ramírez, A., Keizer, C., Van der Sluijs, J. P., Olivier, J., & Brandes, L. (2008). Monte Carlo analysis of uncertainties in the Netherlands greenhouse gas emission inventory for 1990–2004. *Atmospheric Environment*, *42*(35), 8263–8272.
Reuters. (2020). Zara owner offers to make scrubs for Spain's coronavirus-stretched hospitals. Retrieved May 18, 2020, from https://uk.reuters.com/article/us-health-coronavirus-inditex-scrubs/zara-owner-offers-to-make-scrubs-for-spains-coronavirus-stretched-hospitals-idUKKBN2153EK.
Shan, Y., Huang, Q., Guan, D., & Hubacek, K. (2020). China CO_2 emission accounts 2016–2017. *Scientific Data*, *7*, 54.
Shan, Y., Ou, J., Wang, D., Zeng, Z., Zhang, S., Guan, D., & Hubacek, K. (2021). Impacts of COVID-19 and fiscal stimuli on global emissions and the Paris Agreement. *Nature Climate Change*, *11*(3), 200–206.
Shao, S., Liu, J., Geng, Y., Miao, Z., & Yang, Y. (2016). Uncovering driving factors of carbon emissions from China's mining sector. *Applied Energy*, *166*, 200–238.
Shao, S., Zhang, X., & Zhao, X. (2017). Empirical decomposition and peaking pathway of carbon dioxide emissions of China's manufacturing sector—Generalized divisia index method and dynamic scenario analysis. *China Industrial Economics*, *3*, 44–63.
Song, J., Yang, W., Wang, S., Wang, X., Higano, Y., & Fang, K. (2018). Exploring potential pathways towards fossil energy-related GHG emission peak prior to 2030 for China: An integrated input-output simulation model. *Journal of Cleaner Production*, *178*, 688–702.
Tan, X., Lai, H., Gu, B., Zeng, Y., & Li, H. (2018). Carbon emission and abatement potential outlook in China's building sector through 2050. *Energy Policy*, *118*, 429–439.
The State Council of China. (2020). Report on the Work of the Government. http://english.www.gov.cn/premier/news/202005/30/content_WS5ed197f3c6d0b3f0e94990da.html.
Tong, D., Zhang, Q., Zheng, Y., Caldeira, K., Shearer, C., Hong, C., Qin, Y., & Davis, S. (2019). Committed emissions from existing energy infrastructure jeopardize 1.5°C climate target. *Nature*, *572*(7769), 373–377.

UNFCCC. (2015). Adoption of the Paris Agreement. Report No. FCCC/CP/2015/L.9/Rev.1. Available at http://unfccc.int/resource/docs/2015/cop21/eng/l09r01.pdf.

Viebahn, P., Vallentin, D., & Höller, S. (2015). Prospects of carbon capture and storage (CCS) in China's power sector - an integrated assessment. *Applied Energy*, *157*, 229–244.

Vithayasrichareon, P., & MacGill, I. F. (2012). A Monte Carlo based decision-support tool for assessing generation portfolios in future carbon-constrained electricity industries. *Energy Policy*, *41*, 374–392.

Wang, Y., & Liang, S. (2013). Carbon dioxide mitigation target of China in 2020 and key economic sectors. *Energy Policy*, *58*, 90–96.

Wang, H., Lu, X., Deng, Y., Sun, Y., Nielsen, C. P., Liu, Y., Zhu, G., Bu, M., Bi, J., & McElroy, M. B. (2019). China's CO_2 peak before 2030 implied from characteristics and growth of cities. *Nature Sustainability*, *2*(8), 748–754.

Wu, R., Geng, Y., Cui, X., Gao, Z., & Liu, Z. (2019a). Reasons for recent stagnancy of carbon emissions in China's industrial sectors. *Energy*, *172*, 457–466.

Wu, C., Huang, G., Xin, B., & Chen, J. (2018). Scenario analysis of carbon emissions' anti-driving effect on Qingdao's energy structure adjustment with an optimization model, part I: Carbon emissions peak value prediction. *Journal of Cleaner Production*, *172*, 466–474.

Wu, S., Zheng, X., You, C., & Wei, C. (2019b). Household energy consumption in rural China: Historical development, present pattern and policy implication. *Journal of Cleaner Production*, *211*, 981–991.

Xiao, H., Wei, Q., & Wang, H. (2014). Marginal abatement cost and carbon reduction potential outlook of key energy efficiency technologies in China's building sector to 2030. *Energy Policy*, *69*, 92–105.

Yang, J., Hu, H., Tan, T., & Li, J. (2016). China's renewable energy goals by 2050. *Environmental Development*, *20*, 83–90.

Yang, L., Xia, H., Zhang, X., & Yuan, S. (2018). What matters for carbon emissions in regional sectors? A China study of extended STIRPAT model. *Journal of Cleaner Production*, *180*, 595–602.

Ye, R. K., Gao, Z. F., Fang, K., Liu, K. L., & Chen, J. W. (2021). Moving from subsidy stimulation to endogenous development: A system dynamics analysis of China's NEVs in the post-subsidy era. *Technological Forecasting and Social Change*, *168*, 120757.

Yu, S., Zhang, S., Li, X., & Li, L. (2018). China can peak its energy-related carbon emissions before 2025: Evidence from industry. *Energy Economics*, *73*, 91–107.

Yuan, J., Na, C., Lei, Q., Xiong, M., Guo, J., & Hu, Z. (2018). Coal use for power generation in China. *Resources, Conservation and Recycling*, *129*, 443–453.

Yuan, J., Xu, Y., Hu, Z., Zhao, C., Xiong, M., & Guo, J. (2014). Peak energy consumption and CO_2 emissions in China. *Energy Policy*, *68*, 508–523.

Zhang, B., & Chen, G. (2014). China's CH_4 and CO_2 emissions: Bottom-Up estimation and comparative analysis. *Ecological Indicators*, *47*, 112–122.

Zhang, Q., & Fang, K. (2019). Comment on "Consumption-based versus production-base accounting of CO_2 emissions: Is there evidence for carbon leakage?". *Environmental Science Policy*, *101*, 94–96.

Zhang, J., Jiang, H., Liu, G., & Zeng, W. (2018a). A study on the contribution of industrial restructuring to reduction of carbon emissions in China during the five five-year plan periods. *Journal of Cleaner Production*, *176*, 629–635.

Zhang, L., Pang, J., Chen, X., & Lu, Z. (2019). Carbon emissions, energy consumption and economic growth: Evidence from the agricultural sector of China's main grain-producing areas. *Science of the Total Environment*, *665*, 1017–1025.

Zhang, Q., Wiedmann, T., Fang, K., Song, J., He, J., & Chen, X. (2022). Bridging planetary boundaries and spatial heterogeneity in a hybrid approach: A focus on Chinese provinces and industries. *Science of the Total Environment*, *804*, 150179.

Zhang, Q., Xu, J., Wang, Y., Hasanbeigi, A., Zhang, W., & Lu, H., et al. (2018b). Comprehensive assessment of energy conservation and CO_2 emissions mitigation in China's iron and steel industry based on dynamic material flows. *Applied Energy*, *209*, 251–265.

Zhang, X., Zhao, X., Jiang, Z., & Shao, S. (2017). How to achieve the 2030 CO_2 emission-reduction targets for China's industrial sector: Retrospective decomposition and prospective trajectories. *Global Environmental Change*, *44*, 83–97.

Zhou, Y., Hao, F., Meng, W., & Fu, J. (2014). Scenario analysis of energy-based low-carbon development in China. *Journal of Environmental Sciences*, *26*(8), 1631–1640.

Zhou, Y., & Zhao, L. (2016). Impact analysis of the implementation of cleaner production for achieving the low-carbon transition for SMEs in the Inner Mongolian coal industry. *Journal of Cleaner Production*, *127*, 418–424.

5 Carbon Trading Scheme as a Cost-effective Tool for Emission Mitigation

5.1 Kill two birds with one stone: Emission reduction and abatement cost saving

To maintain the increase in world-average temperature well below 2°C, or even below 1.5°C, against pre-industrial levels (UNFCCC, 2010), hundreds of parties have pledged their nationally determined contributions (NDCs) to peak global carbon emissions as early as possible and to pursue the net-zero emissions in the second half of this century (Duan et al., 2021; IPCC, 2014; Pye et al., 2017; Rogelj et al., 2015). In striving to fulfill this ambition, various policy tools such as carbon tax and emission trading scheme (ETS) have come into focus (Mahmood & Marpaung, 2014; Zhang et al., 2019). The former serves as a price instrument that encourages a transition toward cleaner production by imposing extra costs on the emitters proportionally to their emissions (Carattini et al., 2019; Haites, 2018). The latter seems more like a market-based measure with higher flexibility as the carbon price may fluctuate as per the supply and demand of carbon emission allowance (CEA) (Deng et al., 2018; Qian et al., 2018). Moreover, since ETS has the merit of strictly limiting projected emissions below a given level, it has now been widely adopted by the European Union, Switzerland, Australia, New Zealand, South Korean, etc. (Caciagli, 2018), as a critical way of achieving the 1.5°C or 2°C temperature targets (Fang et al., 2021; Pye et al., 2017; Tang et al., 2015).

As the world's largest emitter of greenhouse gases (Lu et al., 2020; Mi et al., 2017a; Wang et al., 2019), China has promised in its NDCs not only to reduce the carbon emissions per unit of GDP (i.e., carbon intensity hereafter) by 60%–65% in 2030 above the 2005 levels but also to peak emissions by latest 2030 (Duan et al., 2018a). To achieve the dual goals in a cost-effective manner, ETS has been implemented by eight pilots in China since 2011 (Cao et al., 2021). While China starts with the establishment of a unified national ETS since 2017, the market is still in its infancy and confronted with some challenges, e.g., inactive trading, steep decline in carbon price, and inadequate incentive for low-carbon investment (Wang et al., 2017a). Moreover, the potential contribution of China's unified ETS to the achievement of the dual goals has yet been thoroughly explored (Jotzo et al., 2018; Weng & Xu, 2018).

In practice, allocating the total CEA to sub-nations is normally a prerequisite for running the national ETS (Fang et al., 2019b; Jotzo & Löschel, 2014; Pang et al.,

DOI: 10.4324/9781003433859-5

2018). As per the Regional Greenhouse Gas Initiative of America, for example, the CEA is first delegated to the states who then take a second allocation to ETS-covered corporations (Ramseur, 2016). China is not an exception, who also delegates the emission reduction targets (which can be converted to CEA) to provinces in its Thirteen Five-Year Plan. However, in most cases, the way how initial CEA can be allocated among regions or corporations has not reached an agreement (Fang et al., 2019b; Liao et al., 2015; Pan et al., 2014). Options include the grandfathering scheme (also referred to as the sovereignty scheme that future CEA for any agents should be proportional to their past emissions) (Mattoo & Subramanian, 2012; Rose et al., 1998), the historical responsibility scheme (referring to that emission reduction tasks should be allocated proportionally to historical cumulative emissions) (Han et al., 2016; Raupach et al., 2014), the ability-to-pay scheme (referring to that the richer agents should take more responsibility for emission reduction) (Wang et al., 2017a; Wu et al., 2016a), and the development right scheme (also referred to as the egalitarian scheme that advocates the equal right of emissions by each person) (Pang et al., 2018; Pye et al., 2017). To assist policymakers in identifying the optimal scheme in practice, evaluating the impact of these initial allocation schemes on the carbon emissions and abatement costs of the country and its provinces would be a much-needed step.

Neglecting the varying marginal abatement costs (MACs) for an additional unit of emission reduction across regions, the regulatory policy tool through CEA allocation may realize the national climate targets at the expense of unaffordable emission abatement costs (EACs). The ETS makes it possible to minimize the total EACs by allowing the agents with different MACs to trade their emission reduction quotas (ERQs) (Liu et al., 2015; Lo, 2012). Figure 5.1 illustrates the mechanism of ETS with a simplified example of two provinces. Suppose that MAC_a and MAC_b are the MAC curves for Provinces *a* and *b* that are entitled to initial reduction targets, $ERQs_a$ and $ERQs_b$. In the absence of the trans-provincial ETS, the equilibrium prices are estimated at P_a and P_b (P_b>P_a) according to the equilibrium points a_0 and

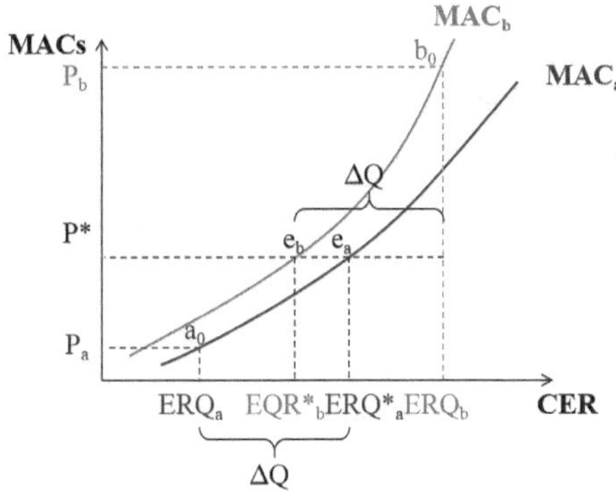

Figure 5.1 Emission trading scheme in terms of marginal abatement cost curves. Abbreviations: CER—Carbon emission reduction; ERQ—Emission reduction quota; MACs—Marginal abatement costs

b_0 where the demand and supply of ERQs are equal. When the trans-provincial ETS is allowed, Province b tends to transfer its ERQs to Province a who has lower MACs, leading to a decrease in the MACs of Province b but an increase in that of Province a. This trading would continue until the MACs of Province a and b are identical at P^* in accordance to the equilibrium points e_a and e_b, and their transferred ERQs are ΔQ.

Implementable methods for designing ETS can be roughly classified into two categories: computable general equilibrium (CGE) models, and programming models (Jiang et al., 2016; Lin & Jia, 2019; Liu et al., 2015). Specifically, CGE models attempt to simulate the activities of various agents, such as residents, corporations, and governments, by connecting several blocks (e.g., the production block, the income-expenditure block) (Dai et al., 2018; Wu et al., 2016b; Yu et al., 2018). Given its capability to fully capture the interaction and feedback effects associated with ETS, this method has won mainstream acceptance over the last decade (Fan et al., 2016; Tang et al., 2016). Nevertheless, CGE models are established on the basis of input–output tables to trace the intermediate commodities along the value chains, and because of this, updates in input–output tables are often slow and can be, in some cases, decades delayed (e.g., the most recent multi-regional input–output tables of Japan are for the year of 1990) (Long et al., 2019, 2020). Furthermore, the complexity of CGE models requires dozens of assumptions that inevitably involve too large uncertainties (Okagawa & Ban, 2008). The constant elasticity of substitution among various commodities, elements, and energy is one example of such a controversial assumption (Lin & Jia, 2017; Wu et al., 2016b).

The programming models that aim for minimizing EACs (or maximizing the profits) have come into focus in simulating ETS. It concentrates on the interaction of agents within the ETS with fewer assumptions and more accessible updated data at the micro levels than CGE. Thus, this approach has increasingly received popularity, particularly in modeling China's ETS aligned with specific policy needs (Chang et al., 2016; Cui et al., 2014; Zhou et al., 2013). Nevertheless, there remain some issues to be addressed in this field. One example is the measurement of the MAC curves, which plays an important role in evaluating the impact of ETS (Cui et al., 2014). Generally, the expert-based model, the CGE model, the distance function model, etc. can be options for this purpose (Kesicki & Ekins, 2012). In this study, we propose a quadratic function for modeling the MAC curves by capturing the correlation between EACs and emission reduction potential. This function is similar in form to existing ones but less data-intensive (Wei & Rose, 2009; Zhou et al., 2013).

A number of studies have investigated the impact of China's ETS on the fulfillment of carbon intensity reduction by 2020 or 2030 while largely neglecting its impact on peaking China's emissions around 2030. However, reaching peak emissions will be more critical to the 1.5°C or 2°C temperature targets (Mi et al., 2017b; Zhang et al., 2016), and much more challenging as it may cause a larger economic loss compared with the goal of carbon intensity reduction (Chai & Xu, 2014; Guan et al., 2018; Yuan et al., 2014). In other words, the contributions of China's ETS can be largely underestimated without consideration of its pathways to peak emissions, which constitutes the research hypothesis to be tested hereafter. Although recently some scholars have attempted to evaluate the contributions of ETS to peaking global emissions (Fujimori et al., 2016; Gallagher et al., 2019), a particular focus on China remains largely lacking, let alone the contributions of ETS to China's dual goals (i.e., to reduce carbon intensity by 60%–65% in 2030, and to peak carbon emissions by latest 2030).

As such, this paper is intended to address an increasingly important yet insufficiently explored dimension of China's climate policy—the potential impact of ETS on China's future emissions and abatement costs in the pursuit of the dual goals as declared in its NDCs by 2030. To that end, we propose a methodological framework aimed at evaluating China's ERQs in line with three scenarios for peak emissions and four schemes for CEA allocation. In doing so, we simulate China's ETS by minimizing its total EACs through nonlinear programming, and provide some valuable insights into the design and operationalization of China's ETS toward the dual goals. The novelty of this paper is summarized as follows: (1) The contributions of China's ETS to the future emissions and abatement costs will be thoroughly explored in terms of its dual goals by 2030, thereby correcting the underestimation under the sole goal of carbon intensity reduction in 2030; (2) Concerning the uncertainty of China's peak emissions, we utilize the scenario analysis to extrapolate various emission trajectories, and distinguish their socio-economic impact; (3) We propose four equity-based schemes and compare their impact on provincial EACs; and (4) Instead of the investment on clean development mechanism that the existing studies focus on, we propose using potential economic loss to fully capture the direct and indirect EACs, which is expected to avoid the distortion of MACs estimation.

The remainder of this paper is structured as follows: Section 5.2 describes the methodology of this paper. The main results are presented in Section 5.3. Section 5.4 compares our findings with literature and discusses the limitations. Section 5.5 includes conclusions and policy recommendations.

5.2 Methods and data

The methodological framework of the research is displayed in Figure 5.2. In this paper, we first evaluate China's overall ERQs by 2030 in three scenarios with different peak-reaching time. This is followed by the initial allocation of provincial ERQs under four schemes for interpreting the equity principle, including the grandfathering, historical responsibility, ability-to-pay, and development right schemes. After measuring the MAC curves for each province, we conduct a nonlinear programming model aimed at minimizing China's total EACs, and simulate China's inter-provincial ETS.

5.2.1 Measurement of China's emission reduction quotas

To measure China's overall ERQs in accordance with the dual goals, the trends in carbon intensity and GDP for the period 2016–2030 should be first anticipated. Following

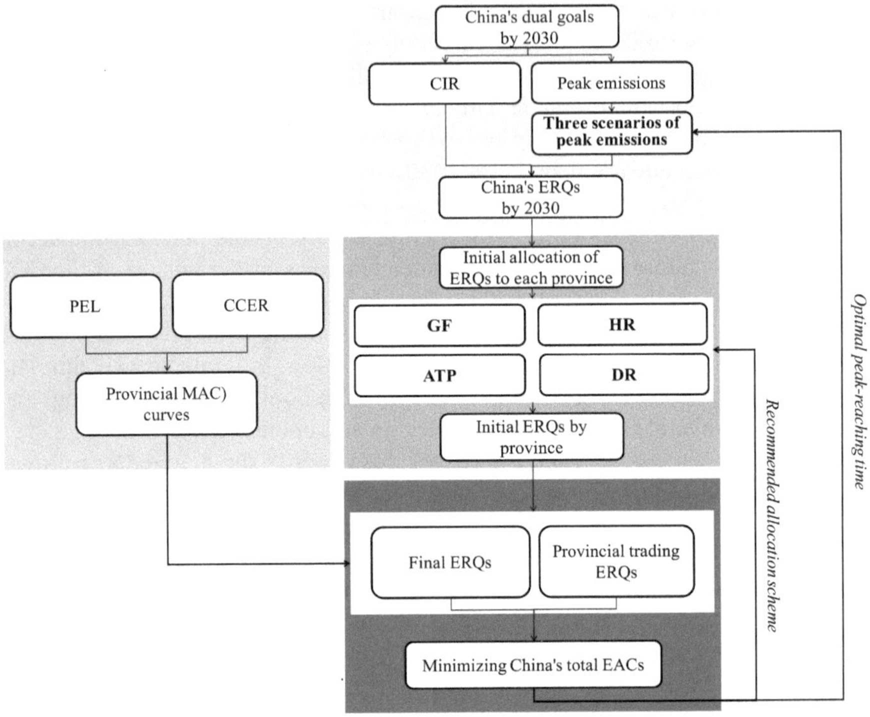

Figure 5.2 The methodological framework of this research. Abbreviations: ATP—Ability-to-pay; CCER—Cumulative carbon emission reduction; DR—Development right; EACs—Emission abatement costs; ERQs—Emission reduction quotas; GF—Grandfathering; HR—Historical responsibility; MAC—Marginal abatement cost

Mi et al. (2017b) and Zhou et al. (2013), this paper assumes an exponential decline in the carbon intensity from 2016 to 2030 at a rate m, following the equations:

$$m = \ln\left(\frac{CI_{2030}}{CI_{2015}}\right)/15 = \ln\left[\frac{CI_{2005} \times (1-65\%)}{CI_{2015}}\right]/15 \tag{5.1}$$

$$CI_t = CI_{2015} \cdot e^{mt} \tag{5.2}$$

where CI_t is the carbon intensity in year t (tons per 10 thousand RMB), and m is the mitigation rate of carbon intensity (%). Admittedly, China's peak emissions are considerably sensitive to its GDP growth rate and peak-reaching time (Pang & Duan, 2016; Weng et al., 2018). To address the uncertainty, we propose three scenarios for China's emissions: peaking in 2020 (Scenario A), in 2025 (Scenario B), and in 2030 (Scenario C). In each scenario, the period 2016–2030 is divided into three time intervals (i.e., 2016–2020, 2021–2025, and 2026–2030), and the corresponding GDP growth rate is assumed in Table 5.1. In order to ensure peak

Table 5.1 The GDP growth rates in the three scenarios

Time intervals	*Time for peaking emissions*		
	2020 (Scenario A)	*2025 (Scenario B)*	*2030 (Scenario C)*
2016–2020	6.50%	6.50%	6.50%
2021–2025	r_1	6.00%	6.00%
2026–2030	r_1	r_1	6.00%

Note: r_1 is the GDP growth rate after peak emissions, which remains constant regardless of the scenarios because its value depends merely on the rate of carbon intensity reduction m that maintains unchanged across the scenarios.

emissions after year t, the GDP growth rate must then be kept below a safe limit expressed as r_1. In other words, it is necessary to maintain a counterbalance between GDP growth and carbon intensity decline after reaching peak emissions. In Scenario B for example, the GDP growth rate before 2025 can be consistent with China's macroeconomic strategies (i.e., The growth rate of GDP is assumed at 6.50% from 2015 to 2020, and at 6.00% from 2021 to 2025, and the uncertainty of economic growth will be discussed later) (National Development and Reform Commission of China, 2012), but it must be adjusted to the level r_1 after 2025 to prevent China's emissions from continuously growing. China's ERQs by 2030 in the three scenarios can be calculated as follows:

$$r_1 = \frac{1}{1+m} - 1 \tag{5.3}$$

$$GDP_t = \begin{cases} GDP_{2015} \cdot e^{r_0 t}, t \le t^* \\ GDP_{2015} \cdot e^{r_0 t^*} \cdot e^{r_1 t}, t > t^* \end{cases} \tag{5.4}$$

$$ERQs_t = \left(CI_{t-1} - CI_t\right) \cdot GDP_t \tag{5.5}$$

where r_0 and r_1 are the GDP growth rate before and after the peak-reaching time t^* (%), respectively, and $ERQs_t$ is China's emission reduction quotas in year t (million tons, Mt).

5.2.2 *Initial allocation of emission reduction quotas to provinces*

To track the potential impact of ETS regarding the initial allocation of CEA, we identify the four prevailing schemes for comparison (Table 5.2), and assume a second-order polynomial function for projecting the provincial carbon emissions from 2000 to 2030:

$$CE_i^t = a_i \cdot t^2 + b_i \cdot t + c_i, \tag{5.6}$$

where CE_i^t is the carbon emissions of province i in year t (Mt), and a, b, and c are the parameters to be estimated. Equations 5.7–5.10 calculate the CEA of province

Table 5.2 Interpretation of the four initial allocation schemes for carbon emission allowance

Initial allocation scheme	*Indicator*	*Interpretation*
Grandfathering scheme	Historical carbon emissions	The CEA for each province should be a proportion of its past carbon emissions.
Historical responsibility scheme	Historical carbon emissions	More historical carbon emissions a province has, less CEA it should be entitled to.
Ability-to-pay scheme	Per capita GDP	Richer provinces should afford more emission reduction tasks than the poorer, so the provincial CEA has a negative correlation with per capita GDP.
Development right scheme	Population	The provincial CEA is allocated proportionally to its population as each person should have equal rights for carbon emissions.

i in year t, expressed as CEA_i^t (Mt), under the four schemes, respectively, following the equations:

$$CEA_i^t = \left(\sum_i CE_i^t - ERQs_t \right) \cdot \frac{CE_i^{2015}}{\sum CE_i^{2015}} \tag{5.7}$$

$$CEA_i^t = \left(\sum_i CE_i^t - ERQs_t \right) \cdot \frac{1/CE_i^{2015}}{\sum 1/CE_i^{2015}} \tag{5.8}$$

$$CEA_i^t = \left(\sum_i CE_i^t - ERQs_t \right) \cdot \frac{1/_{PGDP_i^{2015}}}{\sum 1/_{PGDP_i^{2015}}} \tag{5.9}$$

$$CEA_i^t = \left(\sum_i CE_i^t - ERQs_t \right) \cdot \frac{P_i^{2015}}{\sum P_i^{2015}} \tag{5.10}$$

$$ERQs_i^t = CE_i^t - CEA_i^t \tag{5.11}$$

where $ERQs_i^t$ is the emission reduction quotas of province i in year t (Mt), $PGDP$ is the per capita GDP (RMB), and P is population. Remarkably, some provinces may have negative initial ERQs, which is inconsistent with the policy. For this reason, we reset the negative values as zero, and the initial ERQs for other provinces would be adjusted proportionally.

5.2.3 Measurement of marginal abatement cost curves for provinces

MAC curves dominate the provincial behaviors in such a way that individual provinces determine either to undertake emission reduction themselves or to transact with other provinces. Inspired by Wei and Rose (2009) and Zhou et al. (2013), we measure the MAC curves for each province through a second-order polynomial function. It implies that the costs due to an additional unit of emission reduction tend to increase. In analogy to Zhou et al. (2013), cumulative carbon emission reduction (defined in this paper as the emissions added up between 2005 and the specific year) is adopted as a proxy for the provincial emission reduction potential. Nevertheless, we propose making use of potential economic loss to represent EACs, instead of the investment on clean development mechanism. The latter has experienced a substantial decrease worldwide (including China) since 2008 and therefore may result in a serious distortion in accounting for provincial MACs (Maraseni, 2013; World Bank, 2012). In principle, the index of potential economic loss is expected to capture both direct and indirect costs of emission reduction that occurs along the value chains (Nordhaus, 2017). By multiplying carbon emissions with the difference in emission productivity (defined in this paper as the inverse of carbon intensity) between adjacent years, potential economic loss could be calculated as follows:

$$PEL_i^t = a_i \cdot \left(CCER_i^t\right)^2 + b_i \cdot CCER_i^t + c_i \tag{5.12}$$

$$CER_i^t = \left(CI_i^{t-1} - CI_i^t\right) \cdot GDP_i^t \tag{5.13}$$

$$CCER_i^t = \sum_{t_0}^{t} CER_i^t \tag{5.14}$$

$$PEL_i^t = \left(EP_i^{t-1} - EP_i^t\right) \cdot CE_i^t \tag{5.15}$$

where $CCER_i^t$ is the cumulative carbon emission reduction of province i from starting year t_0 to year t (Mt), while CER_i^t is the carbon emission reduction of province i in year t. In this study, we set 2005 as the starting year for the cumulative carbon emission reduction accounting. PEL_i^t is the potential economic loss of province i in year t (billion RMB), EP_i^t is the emission productivity of province i in year t (10 thousand RMB per ton), expressed as the inverse of provincial carbon intensity, CI_i^t, and a, b, and c are the parameters to be estimated.

5.2.4 Simulation of China's inter-provincial ETS

To simulate China's ETS at the inter-provincial level, we develop a nonlinear programming model aimed at minimizing China's total EACs for the period 2016–2030, following the equations below:

$$min\ TAC = \sum_{i=1}^{30} \int_{CCER_i^{2016}}^{CCER_i^{2030}} PEL_i \, dCCER_i \tag{5.16}$$

$$\text{s.t. } CCER_i^{2030} - CCER_i^{2015} + TRADE_i \geq \sum_{t=2016}^{2030} ERQs_i^t \tag{5.17}$$

$$\sum TRADE_i = 0 \tag{5.18}$$

where *TACs* is China's total EACs (billion RMB), and $TRADE_i$ is the trading ERQs (i.e., sold to or bought from other provinces) by province *i* (Mt). Thus, Equation (5.17) implies that the carbon emissions reduced by province *i* and the trading ERQs should not be less than the initial ERQs allocated to this province. As shown in Equation (5.18), we argue for market clearing, so that the trading volumes of buyers and sellers should be identical during 2016 and 2030.

5.2.5 Data

Data on the national and provincial carbon emissions for this study basically come from China Emission Accounts & Datasets (http://www.ceads.net), and the socio-economic data are derived from Chinese Statistical Yearbooks, as illustrated in Table 5.3. All economic data like national and provincial GDP have been converted into the constant prices in 2005. Due to the lack of access to China's official data on national and provincial carbon intensities, we calculate them by dividing the carbon emissions by their corresponding GDP.

5.3 Results

5.3.1 Overall emission reduction quotas in accordance with China's dual goals

By multiplying the projected carbon intensity until 2030 and GDP in each scenario, China's carbon emission trajectories in keeping with the dual goals in the period 2016–2030 can be explored. Our simulation shows that China's annual emissions in the three scenarios are likely to peak at 13709.20 Mt in 2020, 15821.04 Mt in 2025, and 18258.20 Mt in 2030, respectively. The results prove that the postponement of peak-reaching time could allow for more CEA. For instance, the total CEA for 2016–2030 is estimated at 201.29, 218.06, and 225.23 Gt in Scenarios A, B, and C, respectively (Figure 5.3). However, the later China peaks its emissions, the larger would be the ERQs. Specifically, the overall ERQs in 2016–2030 range from 6050.03 Mt in Scenario A to 6769.76 Mt in Scenario C, and the peak annual ERQs range from 412.06 Mt to 548.79 Mt, respectively.

5.3.2 Initial emission reduction quotas by province under the four schemes

In addition to the variation of China's overall ERQs in the three scenarios, we observe a pronounced difference in the initial ERQs at provincial levels under the

Table 5.3 Data source of this research

Indicators	*Period*	*Data source*
National and provincial carbon emissions	2005–2015	China Emission Accounts & Datasets (CEADs)
National and Provincial GDP	2005–2015	Chinese Statistical Yearbooks
Provincial population	2015	
Provincial per capita GDP		

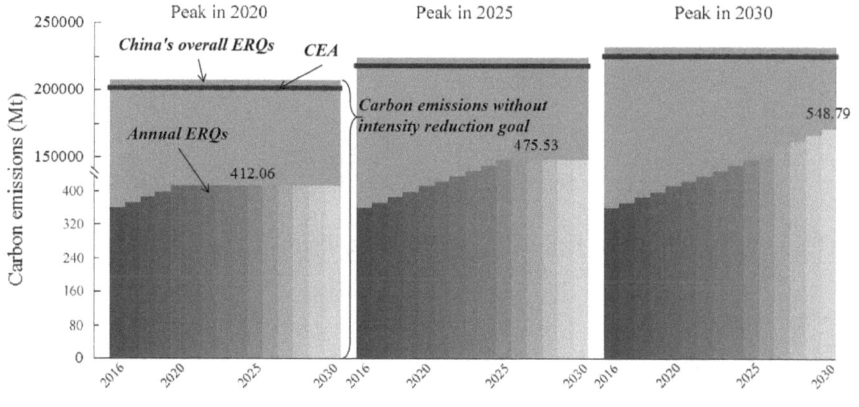

Figure 5.3 China's overall emission reduction quotas (ERQs) from 2016 to 2030

Note: The bottom bars from right to left represent annual ERQs in this time span. The dark lines represent the carbon emission allowance (CEA) in this period with respect to the three scenarios. The upper bars represent the carbon emissions when there is no reduction action for carbon intensity. The difference between the dark lines and upper bars represents China's overall ERQs from 2016 to 2030.

four allocation schemes. Exemplified by Scenario B, Figure 5.4 displays the distribution of initial ERQs by province under each scheme.

Under the grandfathering scheme for example, Shanxi will be entitled to the largest initial ERQs amounting to 2989.45 Mt, followed by Inner Mongolia (1193.58 Mt) and Xinjiang (515.85 Mt). Nine provinces (e.g., Shandong, Jiangxi, Fujian) are exempted from emission reduction because of their low projected carbon emissions and abundant CEA. Under the historical responsibility scheme, 15 provinces are likely to take the initial reduction tasks, where Shanxi and Inner Mongolia still bear the heaviest burdens of initial ERQs (1857.73 and 909.85 Mt), followed by Shandong (627.47 Mt). However, 15 provinces such as Ningxia, Qinghai, and Gansu are not subject to the initial reduction tasks. Under the ability-to-pay scheme, merely 11 provinces have to reduce their emissions, where Shanxi, Inner Mongolia, and Shandong suffer from the largest initial ERQs (2381.10, 1265.82, and 734.73 Mt). With respect to the development right scheme, initial ERQs between 0.71 and 2974.36 Mt would be assigned to 13 provinces (e.g., Shanxi, Inner Mongolia, Shaanxi).

One may extrapolate that some northern and northwestern provinces, such as Shanxi, Inner Mongolia, Shaanxi, and Xinjiang, always suffer from considerably large initial ERQs irrespective of the allocation schemes, whereas Beijing, Shanghai, Fujian, and Jiangxi are four provinces who are not subject to emission reduction in any schemes. Compared to other three schemes, the historical responsibility scheme leads to less difference across all the provinces, as proved by a relatively small standard deviation of provincial initial ERQs (389.92 Mt) in Scenario B. This number rises to 500.62, 576.49, and 611.18 Mt under the ability-to-pay, grandfathering, and development right schemes, respectively.

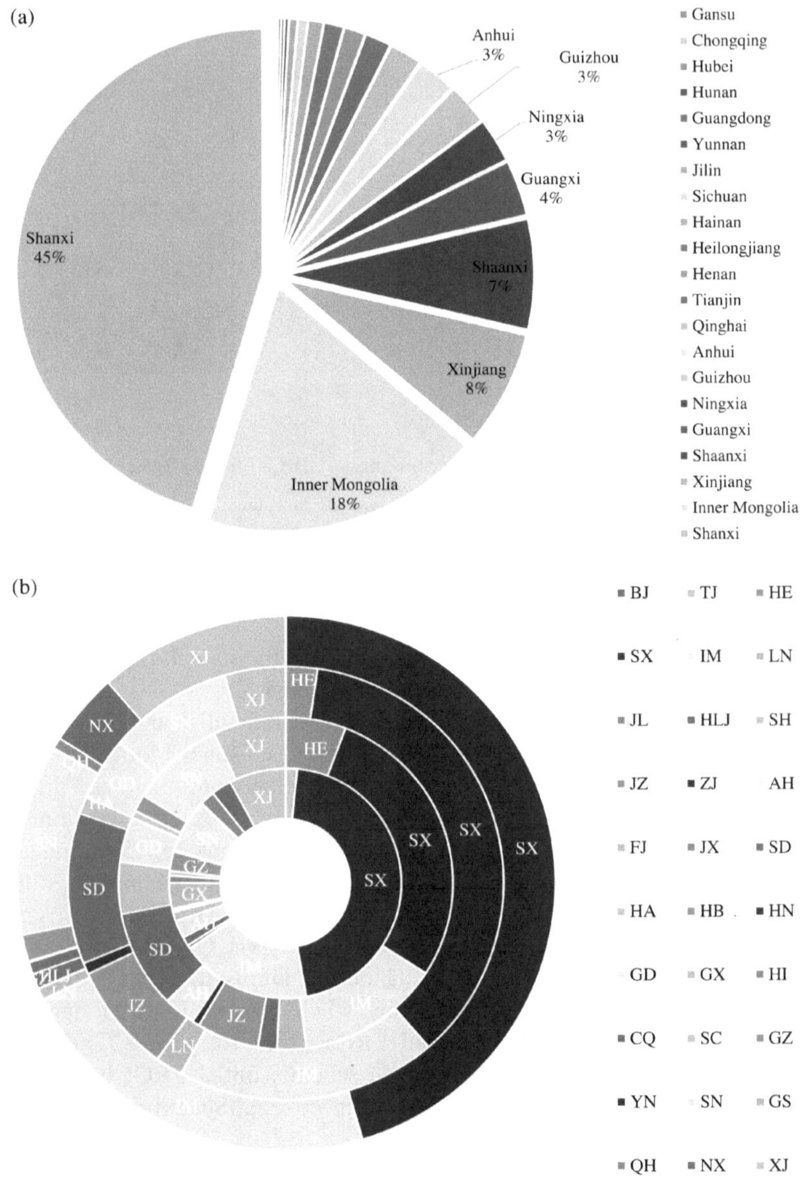

Figure 5.4 The initial emission reduction quotas (ERQs) by province as per the four schemes when China peaks its emissions in 2025 (*Continued*)

Note: The circles from inside to outside represent the distribution in accordance with the grandfathering scheme, the historical responsibility scheme, the ability-to-pay scheme, and the development right scheme. Provinces being entitled to more than 1% of the China's ERQs are marked with names. Abbreviations: BJ—Beijing; TJ—Tianjin; HE—Hebei; SX—Shanxi; IM—Inner Mongolia; LN—Liaoning; JL—Jilin; HLJ—Heilongjiang; SH—Shanghai; JS—Jiangsu; ZJ—Zhejiang; AH—Anhui; FJ—Fujian; JX—Jiangxi; SD—Shandong; HA—Henan; HB—Hubei; HN—Hunan; GD—Guangdong; GX—Guangxi; HI—Hainan; CQ—Chongqing; SC—Sichuan; GZ—Guizhou; YN—Yunnan; SN—Shaanxi; GS—Gansu; QH—Qinghai; NX—Ningxia; XJ—Xinjiang.

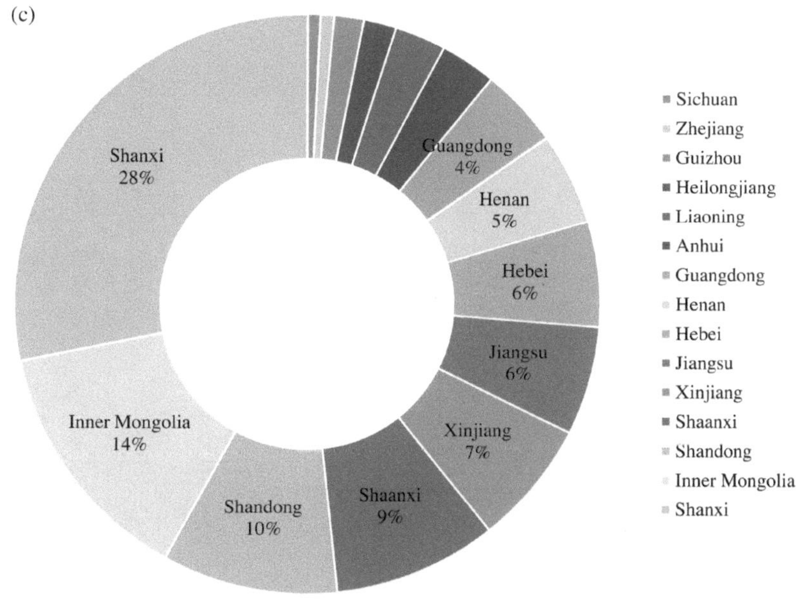

Figure 5.4 (Continued)

5.3.3 Marginal abatement cost curves by province

Figure 5.5 shows the MAC curves of the 30 provinces that are grouped into three regions: the eastern (in red), central (in blue), and western China (in yellow). Figure 5.5(a) and (b) presents the provincial MACs when the cumulative carbon emission reduction is relatively small and large, respectively. The growing provincial CER could lead to an increase in MACs for all provinces. Nonetheless, the trends of MACs vary substantially across different regions. In general, eastern provinces such as Zhejiang, Tianjin, and Beijing would have higher MACs than the majority of central and western provinces. The MACs of central provinces such as Henan, Hubei, and Hunan are higher than those of western provinces (e.g., Chongqing, Sichuan, Guizhou) if the CER is small. For example, when provincial CER is assumed at 100 Mt, the MACs of Hubei and Hunan would exceed 160 RMB per ton CO_2, while those of Yunnan and Shaanxi stay below 120 RMB per ton. However, there is a soar of the MACs in the western provinces if the CER keeps growing. For instance, the MACs of Chongqing, Sichuan, and Guizhou will exceed 3500 RMB per ton when the CER is assumed at 1000 Mt, whereas those of Henan, Hubei, and Hunan are less than 3000 RMB per ton. This demonstrates that most eastern provinces have comparatively less potential for emission reduction as opposed to the central and western ones. Moreover, in most cases where provincial CER is large, western provinces would have basically more difficulty in reducing carbon emissions than central provinces as a result of the convergence of technological levels between eastern, central, and western China.

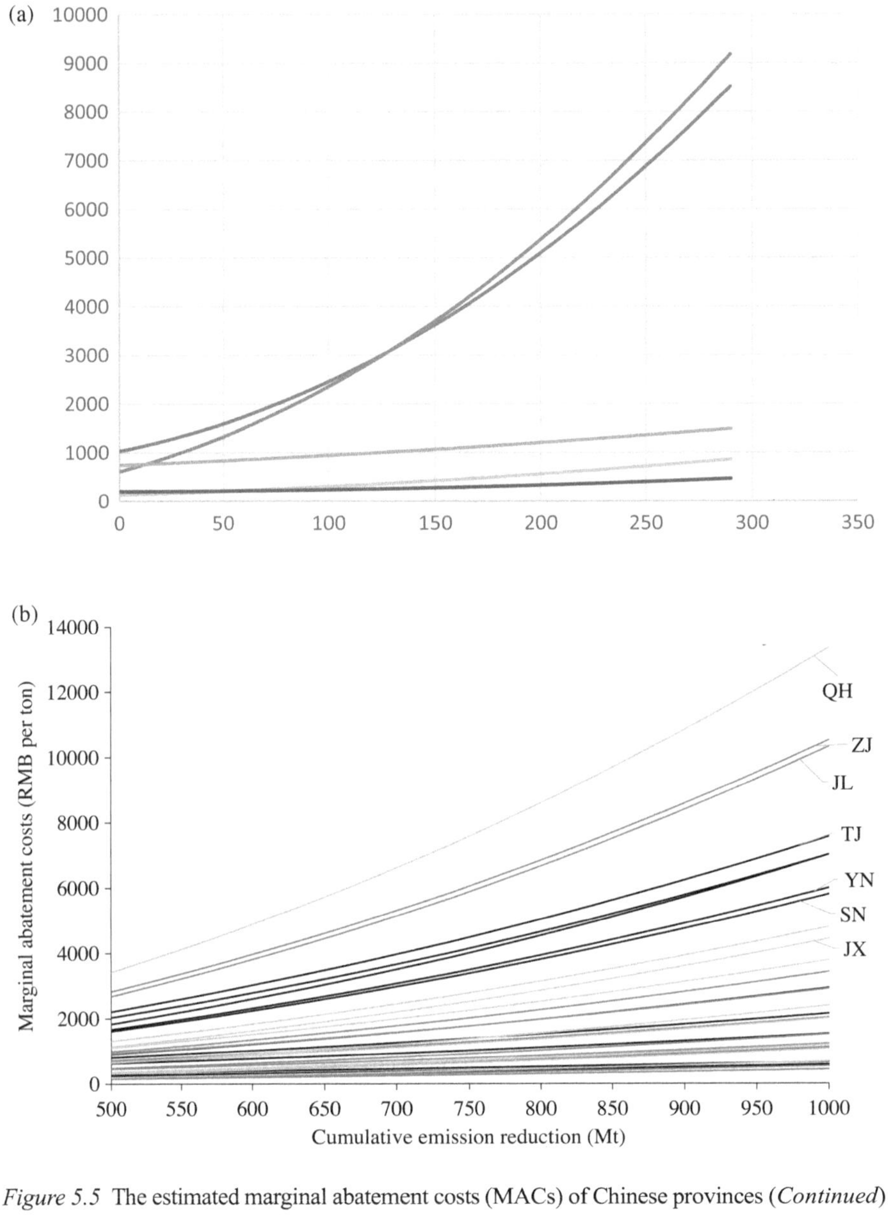

Figure 5.5 The estimated marginal abatement costs (MACs) of Chinese provinces (*Continued*)

5.3.4 Provincial behaviors and benefits in the three scenarios

In Scenario A, China's EACs in the pursuit of the dual goals in 2016–2030 without ETS range from 3491.05 to 6497.13 billion RMB with respect to the four allocation schemes. Remarkably, the total EACs with ETS will make up 0.09% of China's projected GDP during the next 15 years, 60.42%–78.73% less than those

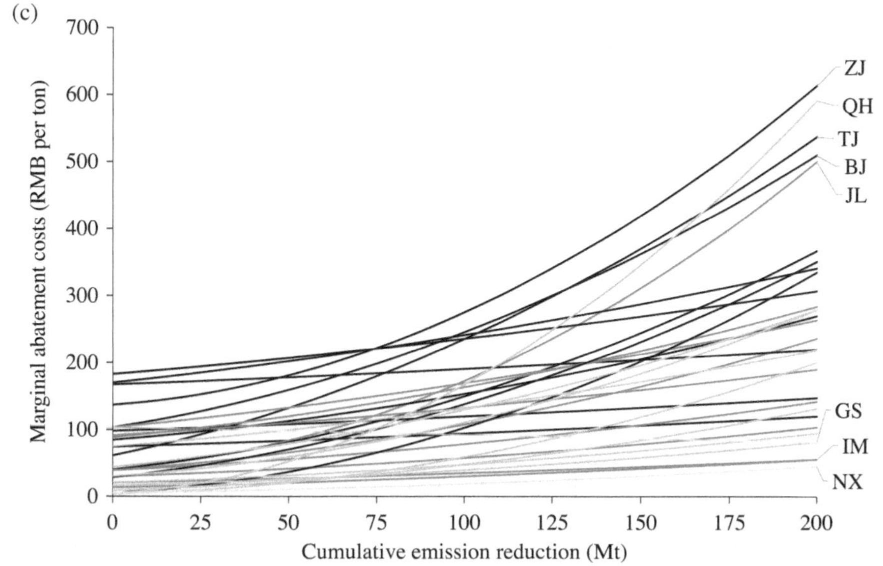

Figure 5.5 (Continued)

without ETS. However, concerning the variation of provincial MACs and their initial ERQs, provinces differ substantially in their behaviors of emission reduction and trading. For example, Figure 5.6(a) displays that 25 provinces are expected to reduce their carbon emissions, as represented by final ERQs (i.e., the initial ERQs minus trading ERQs). Ningxia who has lower MACs will cut its emissions by 654.79 Mt, followed by Inner Mongolia (654.20 Mt) and Hebei (644.52). The costs on final ERQs for these three provinces add up to 114.36, 150.01, and 230.34 billion RMB. On the contrary, the relatively high MACs encourage five provinces increasing their emissions in the period 2016–2030, including Zhejiang (increased by 151.63 Mt), Jiangsu (121.51 Mt), Guangdong (119.19 Mt), Jilin (29.23 Mt), and Tianjin (1.33 Mt), and therefore not having to pay for territorial emission reduction. The remaining 22 provinces would pay 1.01–114.36 billion RMB for their emission reduction.

In Scenario A, the average carbon price is estimated to be 394.72 RMB per ton in 2016–2030. The implementation of national ETS can benefit almost all the provinces, particularly for Shanxi, Shaanxi, and Inner Mongolia despite the initial allocation schemes (Figure 5.6(b)). Hebei is found to be an exception, however, whose total EACs increase by 20.16 billion RMB under the historical responsibility scheme. This partly results from its low MACs and large trading ERQs.

Figure 5.6(c)–(f) shows the trading of provincial ERQs under the four initial allocation schemes. For example, there are 3631.52 Mt ERQs traded inter-provincially under the grandfathering scheme (Figure 5.6(c)). Specifically, ten provinces will sell extra ERQs, such as Shanxi (2181.16 Mt), Inner Mongolia (451.75 Mt), and Shaanxi (177.59 Mt). Conversely, two-thirds of the provinces (e.g., Hebei, Ningxia,

Gansu) act as ERQs buyers whose trading volumes range from 2.44 Mt to 644.52 Mt. Under the historical responsibility scheme (Figure 5.6(d)), the trading of ERQs declines to 3553.09 Mt, where 11 provinces would be the sellers, such as Shanxi (1136.32 Mt), Shandong (583.47 Mt), and Jiangsu (488.28 Mt). The rest 19 provinces act as buyers, including Ningxia (654.79 Mt), Gansu (439.33 Mt), Heilongjiang (312.77 Mt), etc. The ability-to-pay scheme will lead to a large trading ERQs of 4370.13 Mt, where nine provinces (e.g., Shanxi, Shandong, Jiangsu) would sell ERQs amounting to 1.33–1613.36 Mt to others such as Ningxia (654.79 Mt), Hebei (500.79 Mt), and Gansu (439.33 Mt) (Figure 5.6(e)). Under the development right scheme, 4175.18 Mt ERQs would be traded among provinces, for which ten provinces are the sellers, including Shanxi (2172.99 Mt), Inner Mongolia (710.24 Mt), Shaanxi (433.85 Mt), etc. Hebei (643.78 Mt), Gansu (439.33 Mt), Ningxia (387.64 Mt), and other 17 provinces would be the buyers (Figure 5.6(f)).

The postponement of peak-reaching time would definitely increase China's EACs without ETS, amounting to 4111.94–7973.45 billion RMB in Scenario B, and to 4405.53–8671.90 billion RMB in Scenario C, according to the four allocation schemes. Likely, China's ETS implementation will reduce the total EACs to 1605.98 and 1705.55 billion RMB in Scenarios B and C, respectively, both of which make up approximately 0.10% of China's projected GDP for the next 15 years.

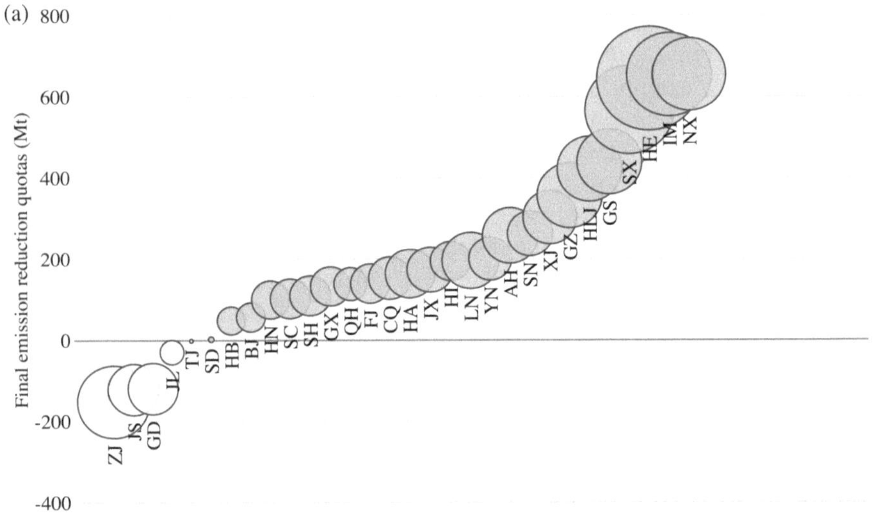

Figure 5.6 Provincial behaviors and benefits when China peaks its emissions in 2020 (*Continued*)

Note: **(a)** Final emission reduction quotas (ERQs) and emission abatement costs (EACs) by province. Negative values imply that those provinces will increase their emissions in the next 15 years. The size of bubbles represents the final EACs. **(b)** Provincial EACs with and without ETS. Provinces below the diagonal will decrease their EACs with ETS, and vice versa. **(c)–(f)** Trading ERQs by province with respect to the grandfathering (GF) scheme **(c)**, the historical responsibility (HR) scheme **(d)**, the ability-to-pay (ATP) scheme **(e)**, and the development right (DR) scheme **(f)**. The positive values in **(c)–(f)** refer to the ERQs sold to other provinces, and vice versa. (*Continued*)

(b)

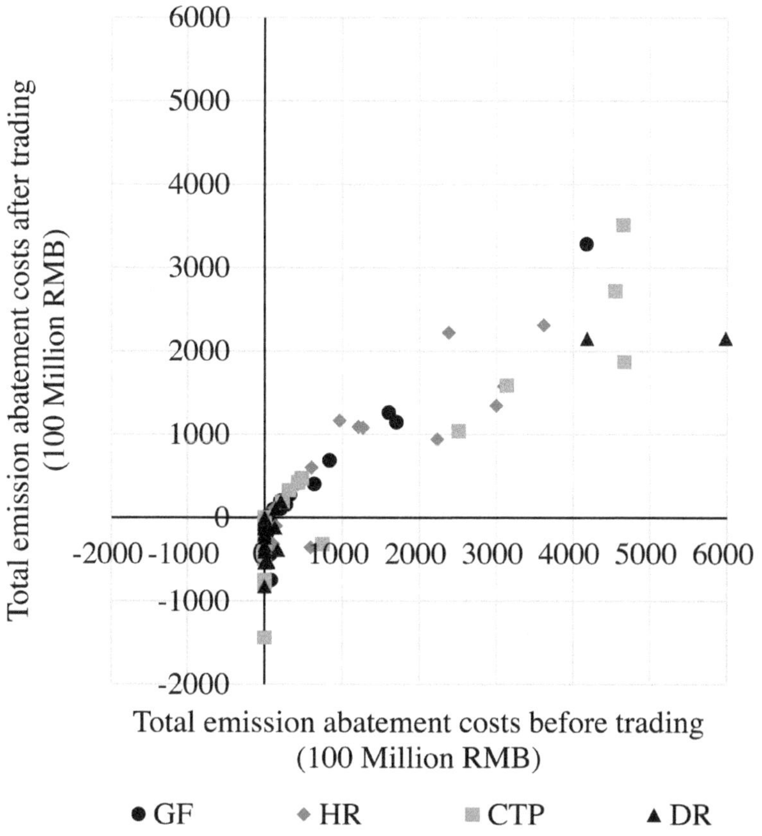

(c)

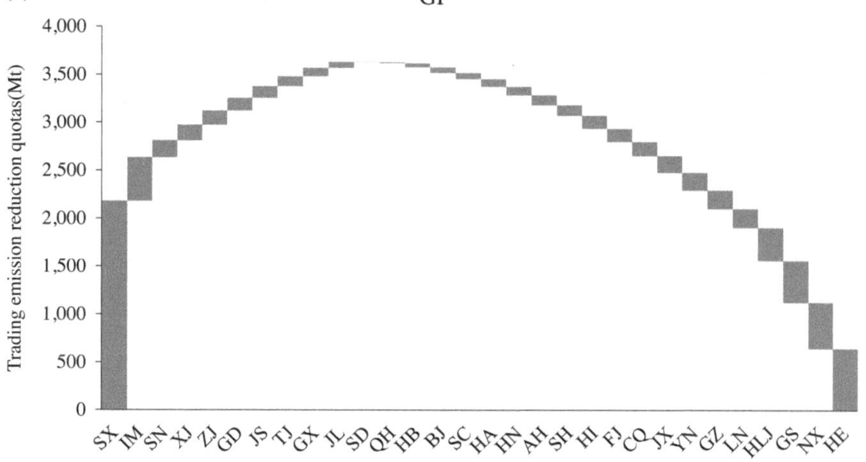

Figure 5.6 (Continued)

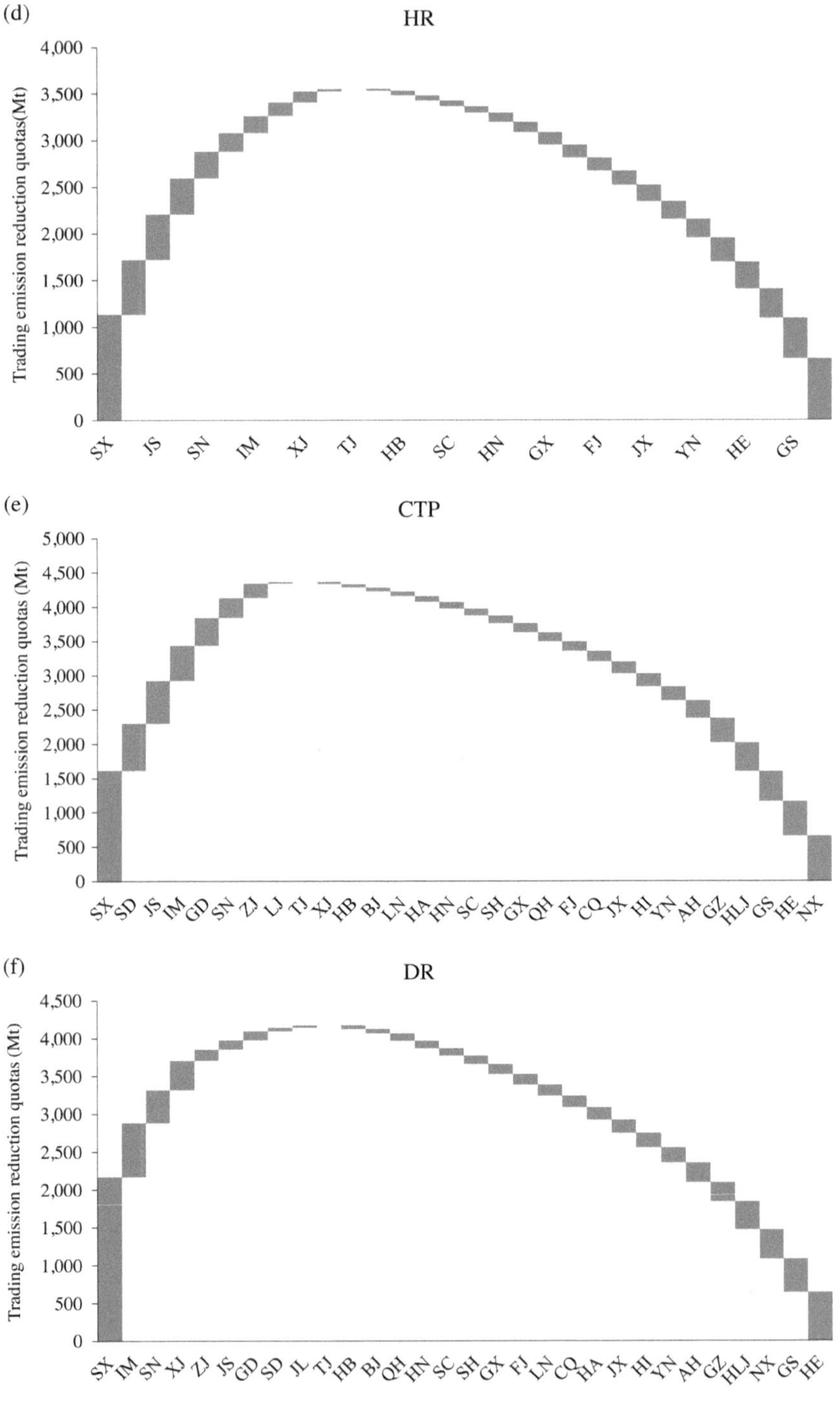

Figure 5.6 (Continued)

Despite the economic benefits brought by ETS at the national level, we notice some difference in the behaviors of individual provinces if peak emissions are postponed, as illustrated in Figure 5.7(a) and Figure 5.8(a). Some northern and northwestern provinces, such as Inner Mongolia, Hebei, and Ningxia, have to substantially reduce their emissions while Zhejiang, Guangdong, Jiangsu, and Jilin are exempted from emission reduction. The largest payment is observed in Hebei for its final ERQs, ranging from 251.64 billion RMB in Scenario B to 261.06 billion RMB in Scenario C, followed by Shanxi, Inner Mongolia, Ningxia, Guizhou, etc. Moreover, the results demonstrate that ERQs would be traded at the price of 416.52 and 426.05 RMB per ton in Scenarios B and C, respectively. In analogy to that in Scenario A, the inter-provincial trading of ERQs through ETS would benefit most of the provinces regardless of the initial allocation schemes, in which Shanxi, Shaanxi, and Inner Mongolia gain the largest benefits from the trading (Figure 5.7(b) and Figure 5.8(b)). However, a growing trend of trading ERQs can be observed in line with the postponement of peak-reaching time (e.g., 3779.41–4658.65 Mt in Scenario B, 3878.49–4784.62 Mt in Scenario C). In general, Shanxi, Inner Mongolia, Jiangsu, and Shaanxi would be the largest sellers of ERQs regardless of the initial allocation schemes, whereas Ningxia, Gansu, and Hebei are the largest buyers.

5.3.5 Economic impact under the four allocation schemes

Figure 5.9(a) compares China's total EACs with and without ETS under the four schemes in each scenario. Our results concretely verify the Coase theorem that the choice of initial allocation schemes has no impact on China's EACs with ETS but can lead to a significant difference in its initial costs. For example, the historical responsibility scheme leads to the lowest EACs without ETS, amounting to 3491.05, 4111.94,

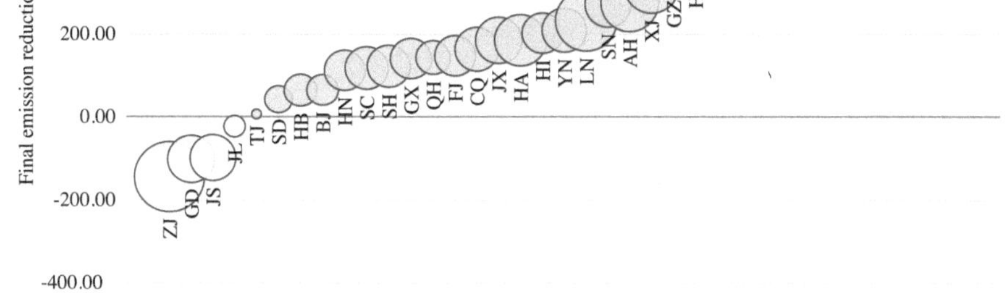

Figure 5.7 Provincial behaviors and benefits when China peaks its emissions in 2025 (*Continued*)

(b)

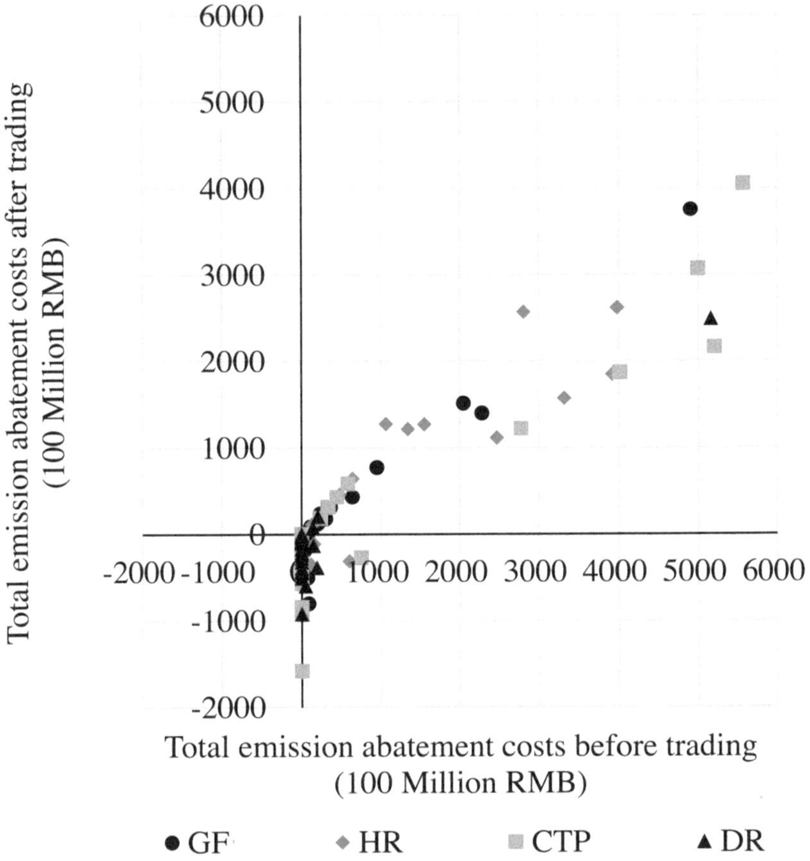

(c)

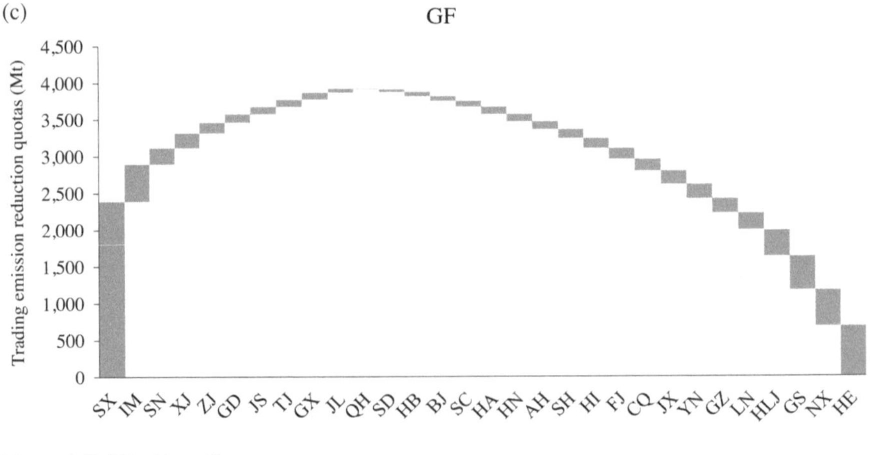

Figure 5.7 (Continued)

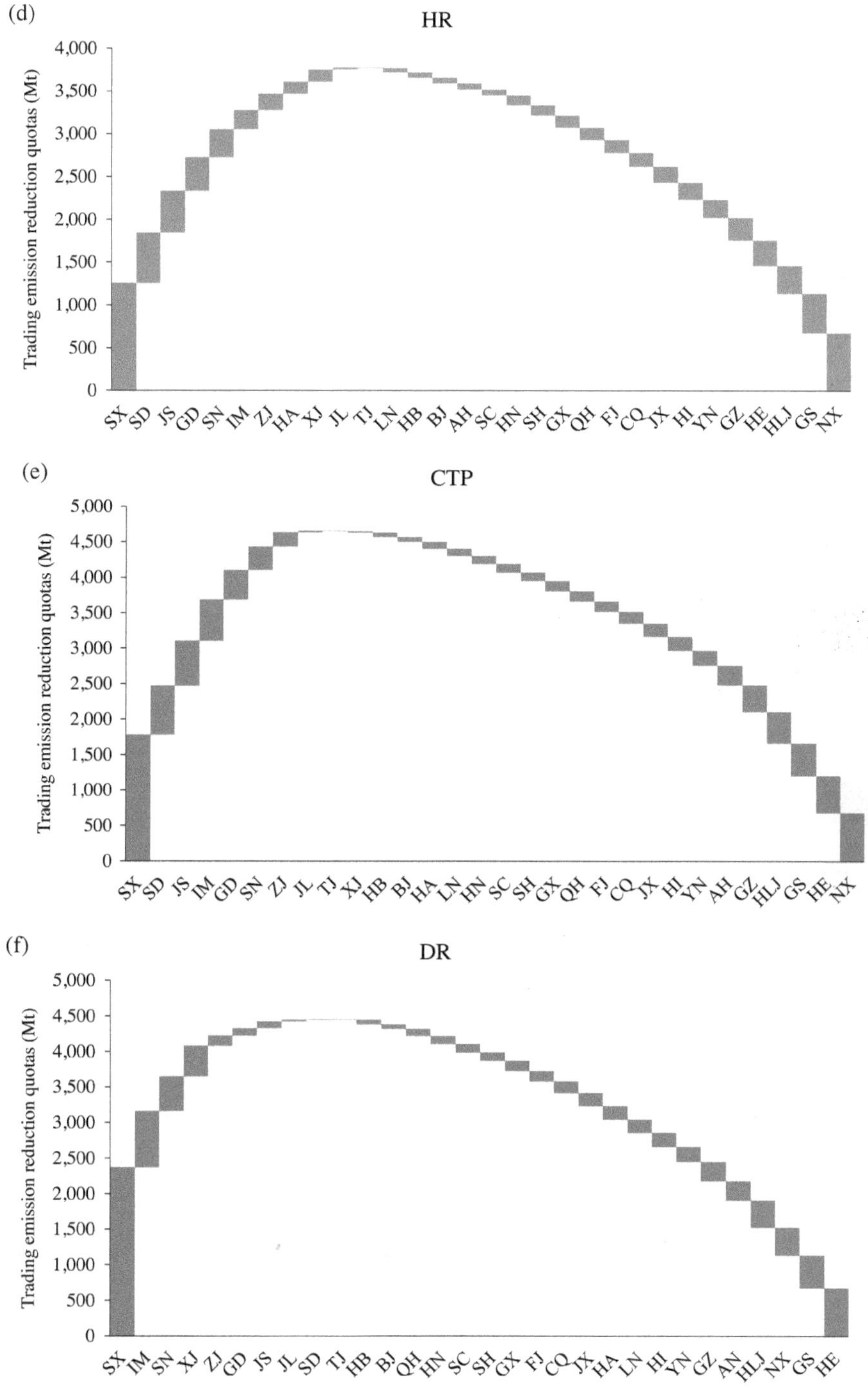

Figure 5.7 (Continued)

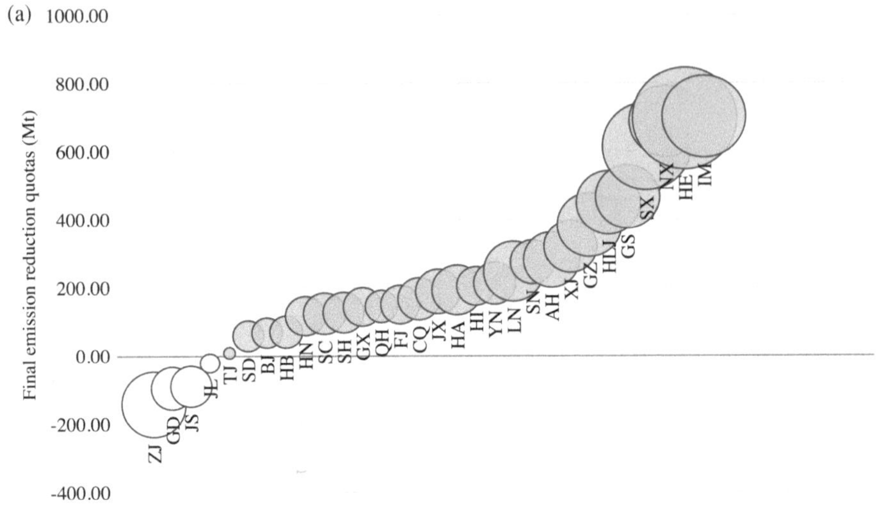

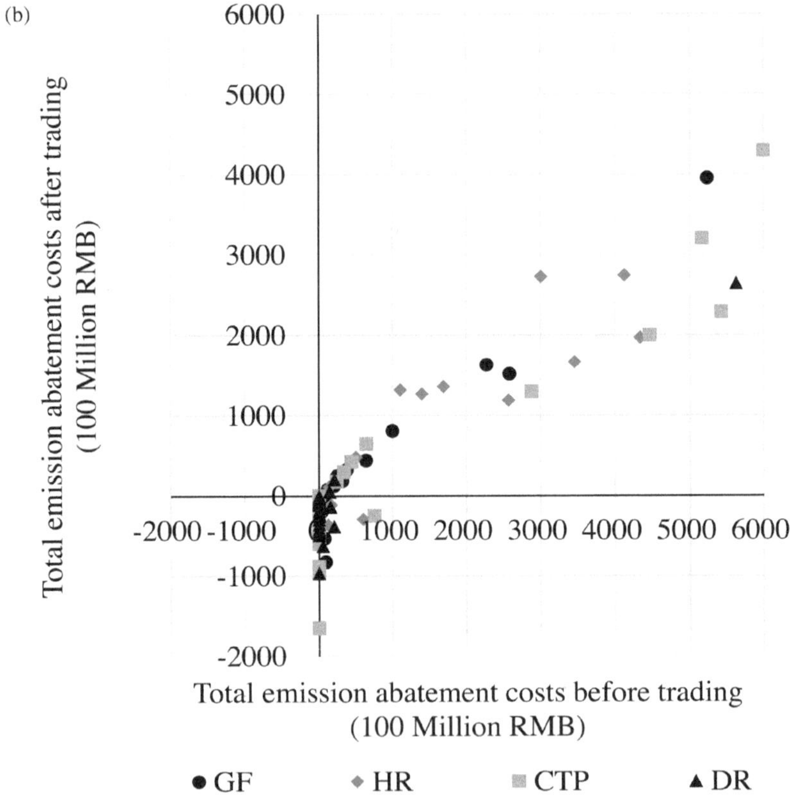

Figure 5.8 Provincial behaviors and benefits when China peaks its emissions in 2030 (*Continued*)

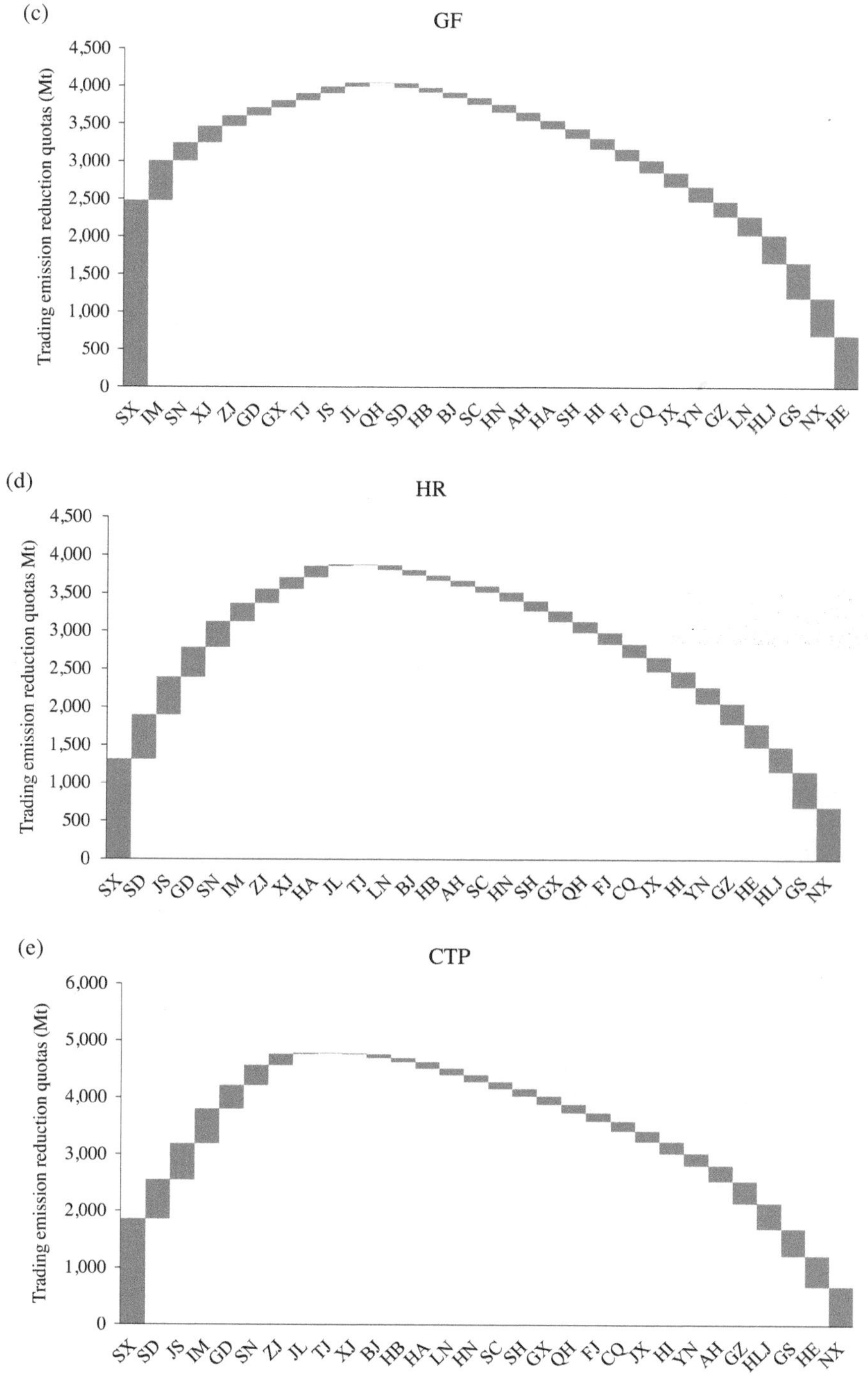

Figure 5.8 (Continued)

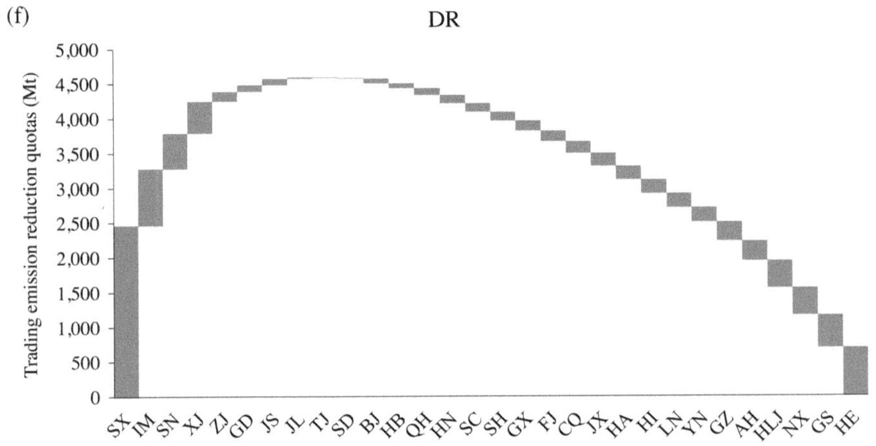

Figure 5.8 (Continued)

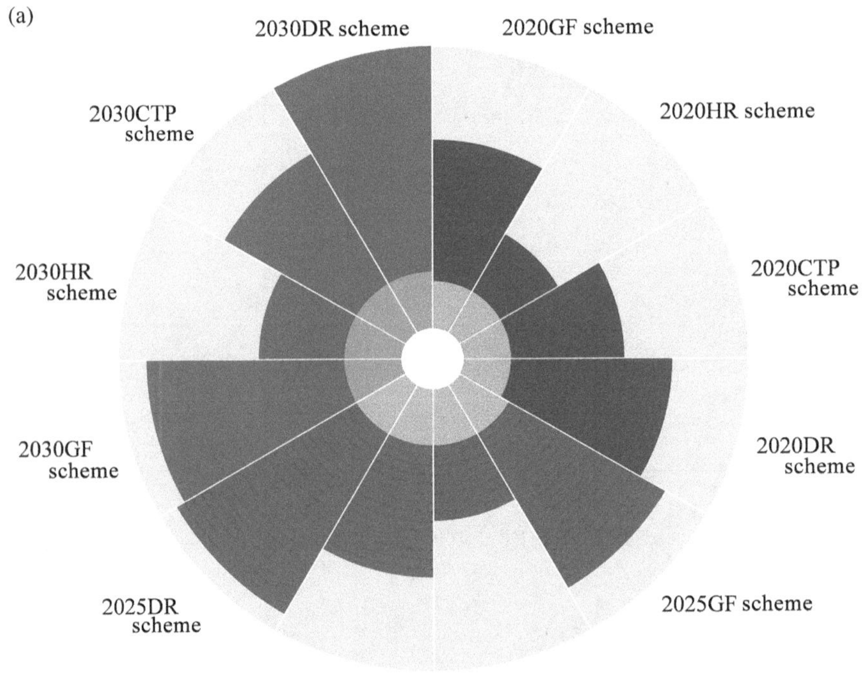

Figure 5.9 China's total emission abatement costs (EACs) and the variation of provincial EACs (*Continued*)

Note: **(a)** China's EACs with and without ETS as marked with light and dark bars, respectively. **(b)** The standard deviation of provincial EACs according to the three scenarios and four schemes. **(c)–(e)** Provincial total EACs under four schemes when carbon emissions peak in 2020, 2025, and 2030, respectively. The dots and colorful areas in (c)–(e) denote the density of provinces concerning their total EACs.

(b)

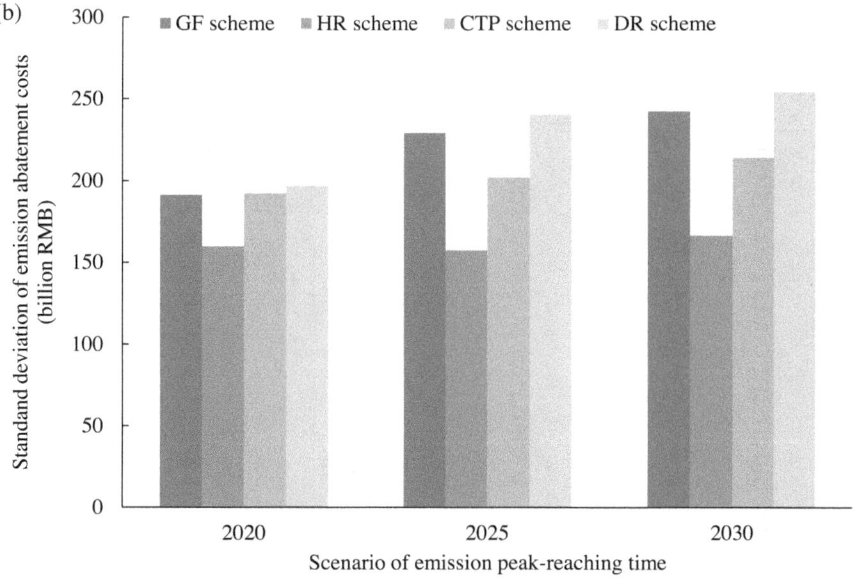

(c)

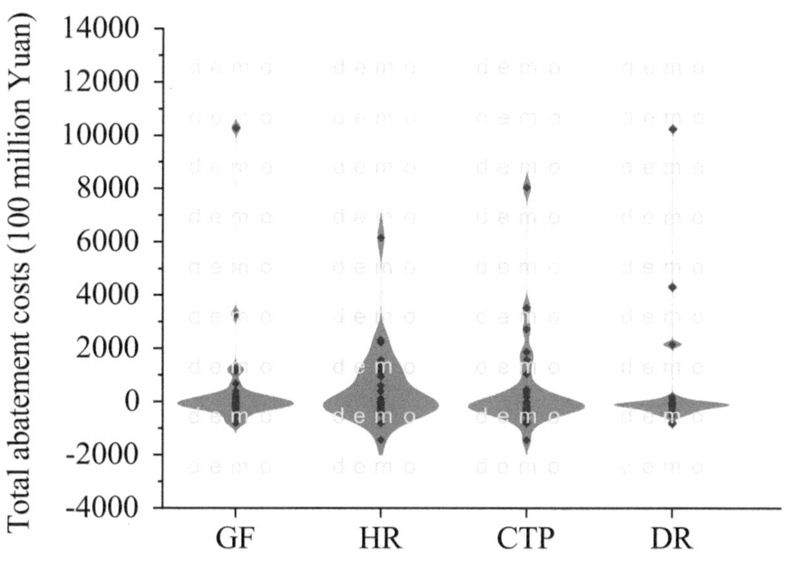

Figure 5.9 (Continued)

(d)

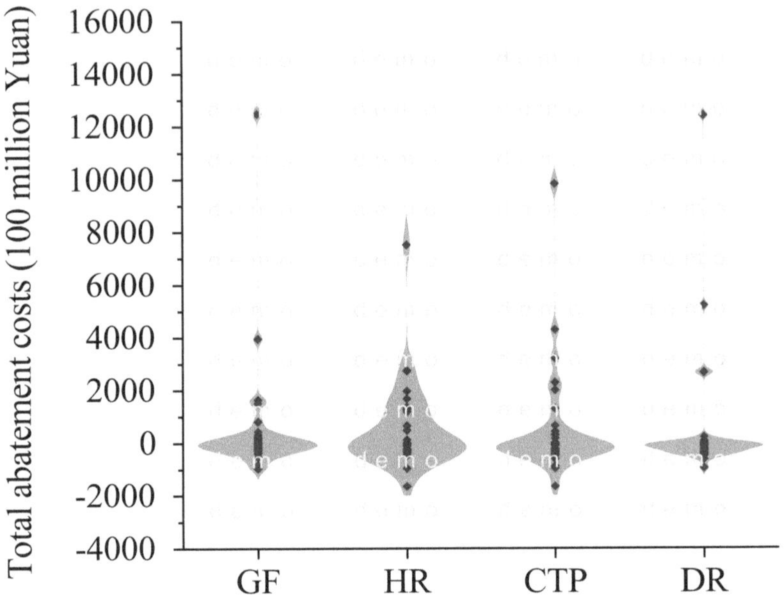

(e)

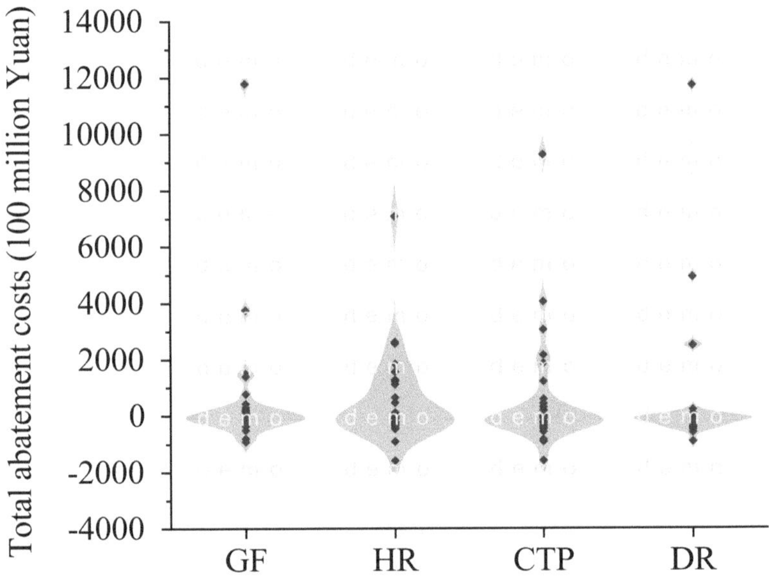

Figure 5.9 (Continued)

and 4405.53 billion RMB in Scenarios A, B, and C. Under the ability-to-pay scheme, the initial EACs for China amount to 4880.73, 5846.38, and 6312.15 billion RMB in Scenarios A, B, and C, respectively. The grandfathering scheme would increase China's initial EACs to 5850.80–7815.77 billion RMB in the three scenarios. Moreover, China's initial EACs would exceed 8000 billion RMB if the development right scheme is adopted. As such, the historical responsibility scheme would be preferred as compared to others because of the lowest initial EACs without ETS.

The results also demonstrate that the some of the initial allocation schemes are likely to make the gap between provincial EACs notably large, perhaps adding obstacles to the implementation of China's ETS in reality. Figure 5.9(b) finds that the historical responsibility scheme could lead to the least variation on provincial EACs despite the scenarios. In Scenario B, for example, the standard deviation of provincial EACs under the historical responsibility scheme is estimated to be 157.28 billion RMB, whereas that under the ability-to-pay, grandfathering, and development right schemes increases to 202.21, 229.12, and 240.92 billion RMB, respectively. For this reason, the historical responsibility scheme is recommended for the initial allocation of CEA to provinces.

5.3.6 Economic impact in the three scenarios

In keeping with the postponement of peak emissions, China's EACs will grow from 1382.00 billion RMB in Scenario A to 1705.54 billion RMB in Scenario C. We also notice that the postponement of peak emissions enlarges the variation of provincial EACs. Under the grandfathering scheme for example, the standard deviation of provincial EACs is estimated at 191.40 billion RMB in Scenario A, and will increase to 229.12 and 242.40 billion RMB in Scenarios B and C, respectively. In this sense, earlier peaking could reduce China's total EACs and contribute to eliminating the variation in provincial EACs.

However, in addition to the two dimensions aforementioned, we must consider China's future GDP in the different scenarios that take a limiting role in determining China's economic growth in the next 15 years. To this end, we evaluate the impact of peak emissions on China's potential GDP in the three scenarios (Figure 5.10). It shows that Scenario C corresponds to the largest potential GDP of 1727892.01 billion RMB for the period 2016–2030. By contrast, the potential GDP in 2016–2030 would be reduced to 1664296.87 and 1525015.68 billion RMB in Scenarios B and A, respectively. In other words, it can trigger some potential GDP loss as compared to Scenario C when the peak emissions are reached earlier. In this sense, there is a dilemma that China's fewer EACs and less gap in provincial EACs (i.e., peaking China's emissions in an earlier time) are obtained at the expense of a larger potential GDP loss. Due to the fact that China's GDP loss considerably outweighs the increase in the national EACs and the standard deviation of provincial EACs, peaking China's emissions in 2030 would be the best choice for making the dual goals come true. Meanwhile, there is a need for employing additional policies (e.g., fiscal transfers, penalty, subsidy) to close the provincial gap that may be posing an obstacle to the operationalization of China's ETS.

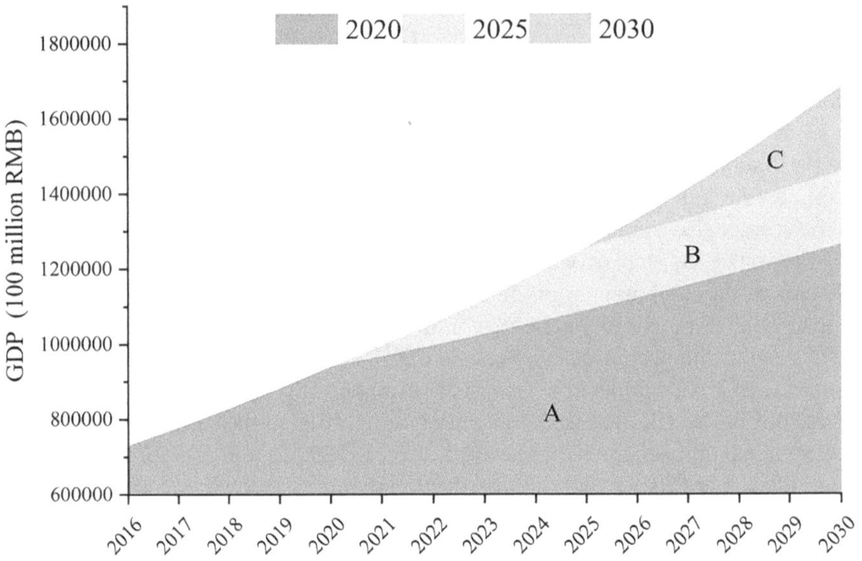

Figure 5.10 The potential GDP in the three scenarios

Note: The area A denotes the projected GDP in 2016–2030 when China's emissions peak in 2020. The area B denotes the increment in projected GDP when China's emissions peak in 2025, and the areas B+C denote the increment in projected GDP when peaking in 2030.

5.4 Discussion

5.4.1 Comparison of the results among scenarios and schemes

Table 5.4 summarizes the changes to main socioeconomic indicators corresponding to the three scenarios and four schemes with or without ETS. As peak emissions are postponed, an ascending trend of China's overall ERQs can be witnessed. Similar trends hold true for China's EACs and carbon price, as well as the standard deviation of provincial EACs. However, we observe an opposite trend for China's potential GDP loss, which is likely to decrease when peak emissions are postponed. While results show a significant variation in China's EACs and the standard deviation of provincial EACs among the four schemes without ETS, it seems clear that the launch of ETS would reduce China's EACs to the same level regardless of the schemes. However, the standard deviation of provincial EACs still differs.

5.4.2 Comparison of the results with other studies

Table 5.5 compares the carbon price and China's EACs for 2030 across studies. Our estimate of China's carbon price ranging from 394.55 to 425.75 RMB per ton is lower than that by Wu et al. (2016b) and Yu et al. (2018), both of which for Shanghai amount to 483.00 and 1030.40 RMB per ton, respectively. Nevertheless, our estimate coincides well with recent literature on China's carbon price by

Table 5.4 Comparison of socioeconomic indicators between different scenarios and schemes with or without ETS

		Time for peaking emissions		
		2020 (Scenario A)	*2025 (Scenario B)*	*2030 (Scenario C)*
China's overall ERQs (Mt)		6050.03	6554.20	6769.76
China's potential GDP/GDP loss (billion RMB)		1525015.68/202876.33	1664296.87/63595.14	1727892.01/0
Carbon price (RMB per ton)		394.72	416.52	426.05
Grandfathering scheme	China's EACs* (billion RMB)	5850.80→1382.00	7182.49→1605.97	7815.77→1705.54
	SD of provincial EACs* (billion RMB)	871.64→191.40	1077.63→229.12	1175.54→242.40
Historical responsibility scheme	China's EACs* (billion RMB)	3491.5→1382.00	4111.94→1605.97	4405.54→1705.54
	SD of provincial EACs* (billion RMB)	285.66→159.73	347.31→157.28	377.57→166.39
Ability-to-pay scheme	China's EACs* (billion RMB)	4880.73→1382.00	5846.38→1605.97	6312.15→1705.54
	SD of provincial* EACs (billion RMB)	504.11→192.35	621.67→202.21	680.16→214.24
Development right scheme	China's EACs* (billion RMB)	6497.13→1382.00	7973.45→1605.97	8671.90→1705.54
	SD of provincial EACs* (billion RMB)	873.91→197.07	1074.54→240.92	1169.58→254.65

Note: '→' represents the change from no-ETS to ETS. Indicators marked with '*' differ with or without ETS, and the rest remain constant. Abbreviation: SD—Standard deviation.

Table 5.5 Comparison of the carbon price and China's emission reduction costs (EACs) in 2030

Reference	*Region*	*Period*	*Carbon price (RMB per ton CO_2)*	*The proportion of EACs in China's potential GDP (%)*
Dai et al. (2018)	China	2017–2030	315.00–819.00	1.20%–2.30%
Lin and Jia (2017)	China	2017–2030	25.20–400.40	0.16%
Wu et al. (2016b)	Shanghai	2030	483.00	2.20%
Yu et al. (2018)	Shanghai	2030	1030.40	3.40%
This paper	China	2016–2030	394.55–425.75	0.10%

Note: US dollar has been exchanged to Chinese RMB with the rate of 7.

Dai et al. (2018) (315–819 RMB per ton) and by Lin and Jia (2017) (25.20–400.40 RMB per ton). Moreover, China's EACs for the period 2016–2030 calculated in this paper are justified by the literature, as its proportion in future GDP (approximately 0.10%) is numerically close to the aforementioned studies (Table 5.5).

The investment on clean development mechanism has experienced a substantial decrease in recent years worldwide and therefore may result in a serious distortion in accounting for provincial MACs. As such, we propose using potential economic loss to avoid such distortion and fully capture the direct and indirect EACs. Many researches provide tangible evidence to support our analysis of the provincial emission reduction and trading behaviors. For example, we find that the largest initial ERQs would be assigned to Shanxi and Inner Mongolia, who tend to sell ERQs for EACs savings, which is consistent with Zhou et al. (2013). By contrast, Hebei, Ningxia, and Gansu would always benefit from reducing additional emissions by buying ERQs because of their lower MACs, which is consistent with Cui et al. (2014).

In addition, our findings highlight the notable contributions of ETS to the EACs savings in realizing the goal of peak emissions apart from carbon intensity reduction by 2030. We show that in the pursuit of the dual goals, the ETS could allow China's total EACs to be saved by 61.41%–80.33%, depending on the specific scenarios and schemes. This appears to be a relatively optimistic estimate by comparison with Yu et al. (2018) at 41.28% and Dai et al. (2018) at 64.45%, probably because we consider not only reducing the carbon intensity but also peaking emissions by 2030. This agrees well with our research hypothesis (as claimed in Section 5.1) that the contributions of China's ETS can be largely underestimated without consideration of the pathways to peak emissions. In that sense, the role of China's ETS in saving EACs can be better captured in the context of the dual goals.

5.4.3 Uncertainties and limitations

Admittedly, this research remains some uncertainties and limitations that should be addressed in future work. First, it is one of the studies depending largely on the assumed parameters like future economic growth that are entangled with uncertainty in China's "New Normal" (Lin, 2011; Weng et al., 2018). However, existing empirical evidence leads us to cautiously conclude that such uncertainty may not be as prominent as previously thought (Cui et al., 2014). Even though the sectoral coverage of ETS can have some notable impact on the EACs of China and all the 30 provinces (Cao et al., 2019; Wang et al., 2017b), the criteria for the inclusion of industrial sectors in ETS have not reached an agreement, thus highlighting the need for further discussions (Mu et al., 2018; Qian et al., 2018). Moreover, while this paper assumes a perfect competition market in which the transaction costs would not be a necessity for modeling the carbon trading, we acknowledge that in practice this may raise some uncertainty on the estimated abatement costs of China and its provinces. However, we believe that this distortion is not unacceptable as we consider the transaction costs in reality and recommend the historical responsibility scheme because of the lower initial abatement costs without ETS. Thus, even though China's ETS cannot reach the ideally equilibrium point as we expect, the historical responsibility scheme can reduce China's abatement costs as much as possible.

Second, we notice that the growth rate of GDP must be adjusted to 3.01% after reaching the peak emissions despite the scenarios to ensure that the dual goals can be fulfilled concurrently. However, this might be somewhat inconsistent with China's macroeconomic strategies. As such, it seems feasible to overfulfill the goal of carbon intensity reduction by 2030, so that more room for economic growth after peaking emissions can be provided. Moreover, recent studies have demonstrated that the overfulfilling of carbon intensity reduction can drive China to an emission trajectory compatible to its latest commitment to net-zero emissions by 2060 (Mallapaty, 2020; Pye et al., 2017). Interestingly, this finding is also consistent with China's latest climate mitigation targets to reduce carbon intensity by no less than 65% in 2030 relative to the 2005 levels, as announced in the Climate Action Summit 2020. But stricter goal of carbon intensity reduction may increase China's total EACs, so future work needs to continuously explore the trade-offs between EACs and economic growth.

Third, our results suggest that a high carbon price would be necessary in support of China's dual goals by 2030. However, such a high price may impede the development of China's ETS market and thus becomes impracticable in short term (Bertram et al., 2015; Fais et al., 2015). Developing the policy package in addition to ETS (e.g., carbon tax, carbon capture and storage deployment, renewable energy subsidy) can somehow make it possible to facilitate the fulfillment of China's dual goals in a more cost-effective way (Li & Taeihagh, 2020; Ye et al., 2021; Yuan et al., 2020). Fourth, the hypothesis of the second-order polynomial relationship between the potential economic loss and CER is somewhat simplified and may add some uncertainty to the final estimate. Although the programming models have been proved to be robust and effective in ETS simulation and the resulting estimates have been justified by literature comparison, it is still desirable to look deeper into the MACs accounting through multiple additional methods.

Last but not least, we assume a free trading of ERQs in the period 2016–2030 across Chinese provinces for brevity. Admittedly, such a long-term market clearing may raise some uncertainty, such as the underestimation of the benefits that some provinces (e.g., Hebei) actually gain. However, many researches demonstrate that the inter-temporal trading can smooth out the demand shocks of ERQs and strengthen the resilience of ETS (Bocklet et al., 2019; Duan et al., 2018b; Fankhauser & Hepburn, 2010). For this reason, the EU allows such an inter-temporal trading through the market stability reserve in its recent reform on ETS (Bocklet et al., 2019; Perino & Willner, 2017). Moreover, although we do not explore how provinces distribute their CEA in each year through a temporal lens, discount rates may have some impact on the carbon emissions and abatement costs of China and its provinces. Dividing the 15-year period into several time intervals is therefore a next step to reduce the uncertainty of such an analysis.

5.5 Conclusions and implications

To capture the contributions of China's ETS to future emissions and abatement costs, this paper goes beyond looking at the goal of carbon intensity reduction by including the simulation of peak emissions by latest 2030. In view of the uncertainty in China's long-term emission trajectories, economic growth, and policy preferences, we explore the potential contributions of ETS to fulfilling the dual goals at three levels: (1) considering with or without ETS; (2) reaching China's peak emissions in 2020, 2025, or 2030; and (3) performing equity-oriented initial allocation of CEA based on the grandfathering, historical responsibility, ability-to-pay, or development right schemes. Major findings include that: (1) China's ETS can largely fuel the dual goals in a cost-effective way where the national total EACs will be reduced by 61.41%–80.33% (i.e., the cost accounts for approximately 0.10% of GDP in 2016–2030), and almost all the provinces could benefit from China's ETS in terms of EACs savings, especially for Shanxi, Shaanxi, and Inner Mongolia; (2) While the postponement of China's peak emissions would lead to an increase in carbon price (from 394.55 to 425.75 RMB per ton), total EACs (from 1382.00 to 1705.54 billion RMB), and the standard deviation of provincial EACs (from 185.14 to 219.42 billion RMB on average), peaking emissions in 2030 is still the preferred choice that will allow for the largest GDP of 1727892.01 billion RMB for the next 15 years; (3) The historical responsibility scheme is preferable to the initial allocation of CEA because it will lead to the least difference in provincial EACs compared with other three schemes; (4) Shanxi, Inner Mongolia, Shaanxi, and Xinjiang would be entitled to the largest initial ERQs regardless of the scenarios and schemes, whereas Beijing, Shanghai, Fujian, and Jiangxi are always exempted from emission reduction; and (5) Shanxi, Inner Mongolia, Jiangsu, and Shaanxi would be the largest ERQs sellers and, conversely, Ningxia, Gansu, and Hebei are the largest buyers despite the scenarios and schemes.

It should be noted that the aim of this paper is to quantify the contributions of China's ETS to achieving the dual goals based on nationwide data that could be of global significance, rather than to demonstrate the rationality and feasibility of these goals which have already been justified by numerous studies (Deng et al., 2018; Gallagher et al., 2019; Wang et al., 2019). Some policy recommendations can be proposed on the basis of our findings: (1) While peaking China's emissions at an earlier time presents a way of somehow reducing China's EACs and the provincial gap, this may cause a considerably larger GDP loss due to the more stringent limit to economic growth. Peaking China's emissions in 2030 would be the best choice for making the dual goals come true. Interestingly, this is consistent with our recent research on peaking simulation from a bottom-up perspective that argues for reaching China's peak emissions in 2030 because of the lowest cumulative carbon emissions between 2016 and 2045 (Fang et al., 2019a); (2) The results verify the Coase theorem that all the four allocation schemes of provincial initial ERQs in question are cost-effective in the context of perfect competition. However, in reality obstacles such as transaction costs and information asymmetry may undermine the efficiency, so the historical responsibility scheme is optimal as it could allow China to have the lowest initial EACs and least provincial difference. Nevertheless, there seems to be room for a combination of different policy tools to eliminate the variation in provincial EACs (e.g., finance transfers, penalty, subsidy); (3) Whichever the scenario applies, China's growth rate of GDP must be adjusted to 3.01% after peaking to prevent emissions from growing. Since this might be somewhat inconsistent with China's macroeconomic strategies, there is a need for overfulfilling the goal of carbon intensity reduction by 2030 to leave more room for economic growth after peaking emissions. This finding is not only consistent with China's updated NDCs but also complies with the latest commitment to achieving net-zero emissions by 2060; and (4) Our simulation that assumes long-term market clearing and constant carbon price may lead to an underestimation of the benefits that ETS can bring. Dividing this long time period into several intervals and examining the differentiated impacts of inter-temporal trading on the agents could be the next step.

References

Bertram, C., Luderer, G., Pietzcker, R. C., Schmid, E., Kriegler, E., & Edenhofer, O. (2015). Complementing carbon prices with technology policies to keep climate targets within reach. *Nature Climate Change*, *5*(3), 235–239.

Bocklet, J., Hintermayer, M., Schmidt, L., & Wildgrube, T. (2019). The reformed EU ETS-intertemporal emission trading with restricted banking. *Energy Economics*, *84*, 104486.

Caciagli, V. (2018). Emission trading schemes and carbon markets in the NDCs: their contribution to the Paris agreement. Theory and practice of climate adaptation, 539–571.

Cao, J., Ho, M. S., Jorgenson, D. W., & Nielsen, C. P. (2019). China's emissions trading system and an ETS-carbon tax hybrid. *Energy Economics*, *81*, 741–753.

Cao, L., Tang, Y., Cai, B., Wu, P., Zhang, Y., Zhang, F., Xin, B., Lv, C., Chen, K., & Fang, K. (2021). Was it better or worse? Simulating the environmental and health impacts of emissions trading scheme in Hubei province, China. *Energy*, *217*, 119427.

Carattini, S., Kallbekken, S., & Orlov, A. (2019). How to win public support for a global carbon tax. *Nature*, *565*, 289–291.

Chai, Q. M., & Xu, H. Q. (2014). Modeling an emissions peak in China around 2030: Synergies or trade-offs between economy, energy and climate security. *Advances in Climate Change Research*, *5*(4), 169–180.

Chang, K., Zhang, C., & Chang, H. (2016). Emissions reduction allocation and economic welfare estimation through interregional emissions trading in China: Evidence from efficiency and equity. *Energy*, *113*, 1125–1135.

Cui, L. B., Fan, Y., Zhu, L., & Bi, Q. H. (2014). How will the emissions trading scheme save cost for achieving China's 2020 carbon intensity reduction target? *Applied Energy*, *136*, 1043–1052.

Dai, H., Xie, Y., Liu, J., & Masui, T. (2018). Aligning renewable energy targets with carbon emissions trading to achieve China's INDCs: A general equilibrium assessment. *Renewable and Sustainable Energy Reviews*, *82*, 4121–4131.

Deng, Z., Li, D., Pang, T., & Duan, M. (2018). Effectiveness of pilot carbon emissions trading systems in China. *Climate Policy*, *18*(8), 992–1011.

Duan, H., Mo, J., Fan, Y., & Wang, S. (2018a). Achieving China's energy and climate policy targets in 2030 under multiple uncertainties. *Energy Economics*, *70*, 45–60.

Duan, M., Qi, S., & Wu, L. (2018b). Designing China's national carbon emissions trading system in a transitional period. *Climate Policy*, *18*(sup 1), 1–6.

Duan, H., Zhou, S., Jiang, K., Bertram, C., Harmsen, M., Kriegler, E., van Vuuren, D., Wang, S., Fujimori, S., Tavoni, M., Ming, X., Keramidas, K., Iyer, G., & Edmonds, J. (2021). Assessing China's efforts to pursue the 1.5°C warming limit. *Science*, *372*, 378–385.

Fais, B., Blesl, M., Fahl, U., & Voß, A. (2015). Analysing the interaction between emission trading and renewable electricity support in TIMES. *Climate Policy*, *15*(3), 355–373.

Fan, Y., Wu, J., Xia, Y., & Liu, J. Y. (2016). How will a nationwide carbon market affect regional economies and efficiency of CO_2 emission reduction in China? *China Economic Review*, *38*, 151–166.

Fang, K., Tang, Y., Zhang, Q., Song, J., Wen, Q., Sun, H., Ji, C., & Xu, A. (2019a). Will China peak its energy-related carbon emissions by 2030? Lessons from 30 Chinese provinces. *Applied Energy*, *255*, 113852.

Fang, K., Wang, S., He, J., Song, J., Fang, C., & Jia, X. (2021). Mapping the environmental footprints of nations partnering the belt and road initiative. *Resources, Conservation and Recycling*, *164*, 105068.

Fang, K., Zhang, Q., Long, Y., Yoshida, Y., Sun, L., Zhang, H., Dou, Y., & Li, S. (2019b). How can China achieve its intended nationally determined contributions by 2030? A multi-criteria allocation of China's carbon emission allowance. *Applied Energy*, *241*, 380–389.

Fankhauser, S., & Hepburn, C. (2010). Designing carbon markets. Part I: Carbon markets in time. *Energy Policy*, *38*(8), 4363–4370.

Fujimori, S., Kubota, I., Dai, H., Takahashi, K., Hasegawa, T., Liu, J. Y., Hijoka, Y., Masui, T., & Takimi, M. (2016). Will international emissions trading help achieve the objectives of the Paris Agreement? *Environmental Research Letters*, *11*(10), 104001.

Gallagher, K. S., Zhang, F., Orvis, R., Rissman, J., & Liu, Q. (2019). Assessing the policy gaps for achieving China's climate targets in the Paris Agreement. *Nature Communications*, *10*(1), 1256.

Guan, D., Meng, J., Reiner, D. M., Zhang, N., Shan, Y., Mi, Z., Shao, S., Liu, Z., Zhang, Q., & Davis, S. J. (2018). Structural decline in China's CO_2 emissions through transitions in industry and energy systems. *Nature Geoscience*, *11*(8), 551–554.

Haites, E. (2018). Carbon taxes and greenhouse gas emissions trading systems: What have we learned? *Climate Policy*, *18*(8), 955–966.

Han, R., Tang, B. J., Fan, J. L., Liu, L. C., & Wei, Y. M. (2016). Integrated weighting approach to carbon emission quotas: An application case of Beijing-Tianjin-Hebei region. *Journal of Cleaner Production*, *131*, 448–459.

IPCC. (2014). IPCC Climate Change 2014: Synthesis Report. Cambridge University Press.

Jiang, J., Xie, D., Ye, B., Shen, B., & Chen, Z. (2016). Research on China's cap-and-trade carbon emission trading scheme: Overview and outlook. *Applied Energy*, *178*, 902–917.

Jotzo, F., Karplus, V., Grubb, M., Löschel, A., Neuhoff, K., Wu, L., & Teng, F. (2018). China's emissions trading takes steps towards big ambitions. *Nature Climate Change*, *8*(4), 265–267.

Jotzo, F., & Löschel, A. (2014). Emissions trading in China: Emerging experiences and international lessons. *Energy Policy*, *75*, 9–16.

Kesicki, F., & Ekins, P. (2012). Marginal abatement cost curves: A call for caution. *Climate Policy*, *12*(2), 219–236.

Li, L., & Taeihagh, A. (2020). An in-depth analysis of the evolution of the policy mix for the sustainable energy transition in China from 1981 to 2020. *Applied Energy*, *263*, 114611.

Liao, Z., Zhu, X., & Shi, J. (2015). Case study on initial allocation of Shanghai carbon emission trading based on Shapley value. *Journal of Cleaner Production*, *103*, 338–344.

Lin, J. Y. (2011). China and the global economy. *China Economic Journal*, *4*(1), 1–14.

Lin, B., & Jia, Z. (2017). The impact of emission trading scheme (ETS) and the choice of coverage industry in ETS: A case study in China. *Applied Energy*, *205*, 1512–1527.

Lin, B., & Jia, Z. (2019). What will China's carbon emission trading market affect with only electricity sector involvement? A CGE based study. *Energy Economics*, *78*, 301–311.

Liu, L., Chen, C., Zhao, Y., & Zhao, E. (2015). China's carbon-emissions trading: Overview, challenges and future. *Renewable and Sustainable Energy Reviews*, *49*, 254–266.

Lo, A. Y. (2012). Carbon emissions trading in China. *Nature Climate Change*, *2*(11), 765–766.

Long, Y., Yoshida, Y., Fang, K., Zhang, H., & Dhondt, M. (2019). City-level household carbon footprint from purchaser point of view by a modified input-output model. *Applied Energy*, *236*, 379–387.

Long, Y., Yoshida, Y., Liu, Q., Zhang, H., Wang, S., & Fang, K. (2020). Comparison of city-level carbon footprint evaluation by applying single-and multi-regional input-output tables. *Journal of Environmental Management*, *260*, 110108.

Lu, Q., Fang, K., Heijungs, R., Feng, K., Li, J., Wen, Q., Li, Y., & Huang, X. (2020). Imbalance and drivers of carbon emissions embodied in trade along the Belt and Road Initiative. *Applied Energy*, *280*, 115934.

Mahmood, A., & Marpaung, C. O. (2014). Carbon pricing and energy efficiency improvement–why to miss the interaction for developing economies? An illustrative CGE based application to the Pakistan case. *Energy Policy*, *67*, 87–103.

Mallapaty, S. (2020). How China could be carbon neutral by mid-century. *Nature*, *586*(7830), 482–483.

Maraseni, T. N. (2013). Selecting A CDM investor in China: A critical analysis. *Energy Policy*, *53*, 484–489.

Mattoo, A., & Subramanian, A. (2012). Equity in climate change: An Analytical review. *World Development*, *40*(6), 1083–1097.

Mi, Z., Meng, J., Guan, D., Shan, Y., Song, M., Wei, Y. M., Liu, Z., & Hubacek, K. (2017a). Chinese CO_2 emission flows have reversed since the global financial crisis. *Nature Communications*, *8*(1), 1712.

Mi, Z., Wei, Y. M., Wang, B., Meng, J., Liu, Z., Shan, Y., Liu, Y., & Guan, D. (2017b). Socioeconomic impact assessment of China's CO_2 emissions peak prior to 2030. *Journal of Cleaner Production*, *142*, 2227–2236.

Mu, Y., Evans, S., Wang, C., & Cai, W. (2018). How will sectoral coverage affect the efficiency of an emissions trading system? A CGE-based case study of China. *Applied Energy*, *227*, 403–414.

National Development and Reform Commission of China, 2012. The second national communication on climate change of the People's Republic of China. https://unfccc.int/resource/docs/natc/chnnc2e.pdf.

Nordhaus, W. D. (2017). Revisiting the social cost of carbon. *Proceedings of the National Academy of Sciences, 114*(7), 1518–1523.
Okagawa, A., & Ban, K.. (2008). Estimation of substitution elasticities for CGE models. Discussion Papers in Economics and Business, 16.
Pan, X., Teng, F., & Wang, G. (2014). Sharing emission space at an equitable basis: Allocation scheme based on the equal cumulative emission per capita principle. *Applied Energy, 113*, 1810–1818.
Pang, T., & Duan, M. (2016). Cap setting and allowance allocation in China's emissions trading pilot programmes: Special issues and innovative solutions. *Climate Policy, 16*(7), 815–835.
Pang, T., Zhou, S., Deng, Z., & Duan, M. (2018). The influence of different allowance allocation methods on China's economic and sectoral development. *Climate Policy, 18*(sup1), 27–44.
Perino, G., & Willner, M. (2017). EU-ETS phase IV: Allowance prices, design choices and the market stability reserve. *Climate Policy, 17*(7), 936–946.
Pye, S., Li, F. G., Price, J., & Fais, B. (2017). Achieving net-zero emissions through the reframing of UK national targets in the post-Paris Agreement era. *Nature Energy, 2*(3), 17024.
Qian, H., Zhou, Y., & Wu, L. (2018). Evaluating various choices of sector coverage in China's national emissions trading system (ETS). *Climate Policy, 18*(sup1), 7–26.
Ramseur, J. L. (2016). *The regional greenhouse gas initiative: Lessons learned and issues for congress*. Library of Congress, Congressional Research Service.
Raupach, M. R., Davis, S. J., Peters, G. P., Andrew, R. M., Canadell, J. G., Ciais, P., Friedlingstein, P., Jotzo, F., van Vuuren, D. P., & Le Quere, C. (2014). Sharing a quota on cumulative carbon emissions. *Nature Climate Change, 4*(10), 873–879.
Rogelj, J., Schaeffer, M., Meinshausen, M., Knutti, R., Alcamo, J., Riahi, K., & Hare, W. (2015). Zero emission targets as long-term global goals for climate protection. *Environmental Research Letters, 10*(10), 105007.
Rose, A., Stevens, B., Edmonds, J., & Wise, M. (1998). International equity and differentiation in global warming policy. *Environmental and Resource Economics, 12*(1), 25–51.
Tang, L., Shi, J., & Bao, Q. (2016). Designing an emissions trading scheme for China with a dynamic computable general equilibrium model. *Energy Policy, 97*, 507–520.
Tang, L., Wu, J., Yu, L., & Bao, Q. (2015). Carbon emissions trading scheme exploration in China: A multi-agent-based model. *Energy Policy, 81*, 152–169.
United Nations Framework Convention on Climate Change (2010). UNFCCC. Outcome of the work of the Ad Hoc Working Group on long-term Cooperative Action under the Convention, Home, Cancun Agreements, Draft decision/CP.16.
Wang, L., Chen, W., Zhang, H., & Ma, D. (2017a). Dynamic equity carbon permit allocation scheme to limit global warming to two degrees. *Mitigation and Adaptation Strategies for Global Change, 22*(4), 609–628.
Wang, H., Lu, X., Deng, Y., Sun, Y., Nielsen, C. P., Liu, Y., Zhu, G., Bu, M., Bi, J., & McElroy, M. B. (2019). China's CO_2 peak before 2030 implied from characteristics and growth of cities. *Nature Sustainability, 2*(8), 748–754.
Wang, X., Teng, F., Zhou, S., & Cai, B. (2017b). Identifying the industrial sectors at risk of carbon leakage in China. *Climate Policy, 17*(4), 443–457.
Wei, D., & Rose, A. (2009). Interregional sharing of energy conservation targets in China: Efficiency and equity. *The Energy Journal, 30*(4), 81–111.
Weng, Q., & Xu, H. (2018). A review of China's carbon trading market. *Renewable and Sustainable Energy Reviews, 91*, 613–619.

Weng, Y., Zhang, D., Lu, L., & Zhang, X. (2018). A general equilibrium analysis of floor prices for China's national carbon emissions trading system. *Climate Policy*, *18*(sup1), 60–70.

World Bank. (2012). State and trends of the carbon market 2012.

Wu, R., Dai, H., Geng, Y., Xie, Y., Masui, T., & Tian, X. (2016b). Achieving China's INDC through carbon cap-and-trade: Insights from Shanghai. *Applied Energy*, *184*, 1114–1122.

Wu, J., Fan, Y., & Xia, Y. (2016a). The economic effects of initial quota allocations on carbon emissions trading in China. *The Energy Journal*, *37*(1), 129–151.

Ye, R. K., Gao, Z. F., Fang, K., Liu, K. L., & Chen, J. W. (2021). Moving from subsidy stimulation to endogenous development: A system dynamics analysis of China's NEVs in the post-subsidy era. *Technological Forecasting and Social Change*, *168*, 120757.

Yu, Z., Geng, Y., Dai, H., Wu, R., Liu, Z., Tian, X., & Bleischwitz, R. (2018). A general equilibrium analysis on the impacts of regional and sectoral emission allowance allocation at carbon trading market. *Journal of Cleaner Production*, *192*, 421–432.

Yuan, Y., Duan, H., & Tsvetanov, T. G. (2020). Synergizing China's energy and carbon mitigation goals: General equilibrium modeling and policy assessment. *Energy Economics*, *89*, 104787.

Yuan, J., Xu, Y., Hu, Z., Zhao, C., Xiong, M., & Guo, J. (2014). Peak energy consumption and CO_2 emission in China. *Energy Policy*, *68*, 508–523.

Zhang, D., Zhang, Q., Qi, S., Huang, J., Karplus, V. J., & Zhang, X. (2019). Integrity of firms' emissions reporting in China's early carbon markets. *Nature Climate Change*, *9*(2), 164–169.

Zhang, X., Karplus, V. J., Qi, T., Zhang, D., & He, J. (2016). Carbon emissions in China: How far can new efforts bend the curve? *Energy Economics*, *54*, 388–395.

Zhou, P., Zhang, L., Zhou, D. Q., & Xia, W. J. (2013). Modeling economic performance of interprovincial CO_2 emission reduction quota trading in China. *Applied Energy*, *112*, 1518–1528.

6 Common but Differentiated Responsibility for Emission Mitigation

6.1 Debates on schemes for sharing carbon emission allowance

Numerous studies have observed a near-linear relation between global warming and carbon dioxide (CO_2) emissions associated with human activities since the pre-industrial era (IPCC, 2015; Raupach et al., 2014; Tokarska et al., 2016). In striving to combat climate change through collaborative initiatives worldwide, more than 190 parties (nations and regions) to the United Nations Framework Convention on Climate Change (UNFCCC) formally reached an agreement in Paris in December 2015 that the global-average temperature rise should stay below 2°C, preferably below 1.5°C, above pre-industrial levels (UNFCCC, 2010). The Paris Agreement on climate change represents a landmark in global environmental governance not only because it legitimates the agreed threshold level (<2°C) for increase in temperature but also because it differs in nature from previously documented "top-down" international agreements, such as the Kyoto Protocol, in the sense that it leaves more room for individual parties to formulate their Intended Nationally Determined Contributions (INDCs). As the top-down approach has made little progress and is now at an impasse (Grasso & Roberts, 2014; Raupach et al., 2014), there is a growing interest in assessing the effects of "bottom-up" INDCs on adaptation and mitigation of climate change (Hof et al., 2017; Rogelj et al., 2016).

As the largest CO_2 emitter in the world (EIA, 2009), China played a constructive role in the negotiation of the Paris Agreement by submitting its ambitious INDCs (Chu, 2017; Ye et al. (2017). The overarching target of China's INDCs is to reduce the intensity of CO_2 emissions by 60–65% in 2030 as compared to the level of 2005. It represents a step ahead of the goal announced in 2009 to reduce the CO_2 emission intensity by 40–45% between 2005 and 2020 (Guan & Delman, 2017; Shao et al., 2016). Meanwhile, the Chinese economy has now entered "New Normal"—a new phase focusing more on quality than on speed (Lin, 2011). A reasonable expectation for China's economic growth between 2020 and 2030 would be at an average rate of 4–6%, depending on the outcome of implementing structural reforms (Green & Stern, 2015). In that sense, future allowable carbon emissions for China will become a finite common-pool resource that must be shared among different parts of the country, especially at the provincial level (Ju & Fujikawa, 2019). This

DOI: 10.4324/9781003433859-6

brings into focus the provincial allocation of China's carbon emission allowance (CEA) in keeping with INDCs.

While to our knowledge research into this question remains underexplored, lessons can be learned from existing literature seeking to determine the CEA by province in China in accordance to the goal for carbon emission control by 2020. For instance, a composite index was proposed by Yi et al. (2011), who allocated China's CEA at the provincial level by 2020 by aggregating GDP per capita, accumulated carbon emissions, and energy consumption per unit of industrial added value under four scenarios. In a similar study, Wang et al. (2011) brought together per capita GDP, per capita emissions, energy consumption of industry per unit of value-added, and the ratio of non-fossil fuel to primary energy consumption into a so-called China Regional Burden Differentiation Model to determine China's provincial CEA in 2020. Yu et al. (2014) measured the contribution of 13 factors to the carbon emissions of 30 Chinese provinces and then allocated the national CEA to each province on the basis of the anticipated growth rate of each factor in 2020. By making use of the Shapley value method to China's CEA allocation for 2020, Zhang et al. (2014) concluded that regions with indicators like higher GDP and carbon outflow should be allocated with more emission quotas.

To operationalize the CEA allocation, some researchers have chosen to focus on the cities as a unit of analysis. Han et al. (2016) developed a similar system of indicators through an analytic hierarchy process method and accounted for the CEA of all the cities within Jing-Jin-Ji region in keeping with China's 2020 goal. Li et al. (2018b) applied a maximum deviation method to allocate the CEA increment from 2015 to 2020 to different Chinese cities in the Pearl River Delta region by taking into account population, GDP, and historical carbon emissions. Despite the differing allocation schemes employed in these studies aforementioned, most of them consider the equity principle as the first principle of distributive justice. The significance of the equity principle to sharing climate burdens and emission rights has been demonstrated in an extensive literature, from diverse perspectives such as egalitarian (per capita emissions), ability-to-pay (per capita GDP), grandfathering (historical carbon emissions), etc. (Pan, 2003; Pan et al., 2014; Wang et al., 2017).

At the same time, the need for multicriteria allocation schemes has been noticed. A prominent advantage over single-criterion allocation is that multicriteria allocation is likely to give rise to less difference between the smallest and largest targets for different entities, which enables more consensus-based entitlements (Ringius et al.,1998). Moreover, multicriteria allocation schemes make it possible to bridge the gap between developed and developing countries who often give preferences to different allocation principles (Wang et al., (2017). The inclusion of various allocation principles allows for the consideration of disparity among provinces and ultimately facilitates the implementation of the CEA allocation scheme. Thus, in the search of implementable allocation schemes for CEA, additional allocation principles such as efficiency, feasibility, and sustainability have been increasingly adopted beyond the equity principle (Fang et al., 2018a; Wang et al., 2011; Yi et al., 2011). On the other hand, the challenge for mitigation and adaptation of climate change in different regions would depend primarily on their respective social, economic, and environmental status (Kriegler et al., 2010; Marangoni et al., 2017). In that sense, one-to-one correspondence can be identified between the different principles (i.e., equity, efficiency, feasibility, and sustainability) and sustainability pillars (i.e., social, economic, and environmental). For instance, the feasibility principle can be interpreted as the abatement cost of carbon emissions from the economic pillar (Wang et al., 2011), and the sustainability principle from the environmental pillar could refer to the absorptive capacity to sequestrate CO_2 (Fang et al., 2018b).

In developing multicriteria allocation schemes, various methods have been introduced. Examples include the single indicator approach, composite index approach, and game theoretic approach (Zhou & Wang, 2016). Of these, the composite index approach aimed at bringing together diverse principles and indicators with determined weights has reached worldwide popularity (Han et al., 2016; Zhou & Wang, 2016). However, the approach has been criticized for its subjectivity and arbitrariness, but also for the fact that it focuses on the absolute amount of indicators (e.g., provincial GDP, population) while looking down upon the relative performance of the inflows and outflows in the whole system (Chiu et al., 2013; Miao et al., 2016; Zhang & Hao, 2017). As such, some provinces might obtain redundant CEA compared with its actual needs, and vice versa.

By contrast, an optimization approach aimed at improving the technical efficiency of the whole system, namely data envelopment analysis (DEA), has emerged as another common way of allocating CEA. The traditional DEA model was proposed by Charnes et al. (1978) and Banker et al. (1984), assuming that all decision-making units (DMUs) in a free market can freely produce input and output variables and that the DEA efficiency of DMUs can be measured through the ratio of multiple inputs to outputs. However, this hypothesis does not hold true when it comes to CEA allocation, as in this context the overall CEA of a given region should be limited to a certain range of values. To overcome this weakness, a revised DEA model, namely zero sum gains-DEA (ZSG-DEA), was proposed by Lins et al. (2003). The rationale of ZSG-DEA for allocating CEA is that when a DMU (e.g., a city, a province, a country, etc.) decreases its CEA to yield greater DEA efficiency, the CEA entitled to any other DMUs must be increased by the same amount to maintain the overall CEA unchanged. By taking the relative performance of different DMUs into consideration without weighting (as opposed to the composite index approach), DEA or ZSG-DEA method allows for a reallocation of the redundant CEA among provinces given their individual technical efficiency.

By adopting population and energy consumption as input variables, and GDP and carbon emissions as output variables, Gomes and Lins (2008) pioneered to allocate the CEA among the UNFCCC Annex I parties utilizing a ZSG-DEA model in order to achieve the goal of Kyoto Protocol. Afterwards, an increasing number of studies have implemented ZSG-DEA as a means to ensure a robust and reasonable allocation of resources and emission allowance. For instance, Wang et al. (2013) introduced a ZSG-DEA model to allocate China's CEA by 2020 toward 30 provinces by selecting total energy consumption, CO_2 emissions and non-fossil energy consumption as inputs, and GDP and population as outputs. By creating a

ZSG-DEA model that set population and energy consumption as inputs and GDP and carbon emissions as outputs, Pang et al. (2015) redistributed the CEA among 124 countries that were subject to the Kyoto Protocol in 2010. Similarly, Chiu et al. (2015) evaluated the DEA efficiency of 24 EU nations from 2005 to 2007 on the basis of a ZSG-DEA model where national CEA was an input variable and energy consumption, government spending and GDP were output variables. Wen and Zhang (2016) developed a non-radial ZSG-DEA model in support of the allocation of CEA across the 30 Chinese provinces in 2020 by treating carbon emissions and GDP as outputs, and capital stock, population, and energy consumption as inputs. Likewise, Miao et al. (2016) took advantage of a ZSG-DEA model to investigate the DEA efficiency of provincial carbon emissions in China from 2006 to 2010 by choosing a capital stock, population, and energy consumption as inputs, and GDP and carbon emissions as outputs.

Irrespective of the prevalence of ZSG-DEA models for determining the CEA, particularly at the provincial level, it remains problematic not only due to the lack of transparency in defining allocation principles other than the equity and efficiency principles but also in selecting indicators as a basis for quantifying the various principles and reflecting regional differences in a systematic way. Moreover, when it comes to running a ZSG-DEA model, the boundary between input and output variables often seems to be ambiguous. For instance, carbon emissions in some studies are modeled as inputs while in some others they are outputs. Besides, as noted above, existing literature focuses primarily on the way to achieve the China's 2020 goals; therefore, there is a great need for aligning timely provincial allocation schemes with China's INDCs toward 2030.

To promote the transparency, scientific robustness, and policy relevance of CEA allocation, this paper is intended to introduce an improved ZSG-DEA model for the allocation of provincial CEA in accordance to China's INDCs responding to the Paris Agreement. Overall, the novel contributions of this article to the literature lie in (1) integrating multicriteria allocation principles and indicators into an improved ZSG-DEA model; (2) developing an optimal scheme for the allocation of CEA to achieve China's INDCs at the provincial level; (3) serving as a reference for determining the fair share of resources or emission permits across regions; and (4) providing policy recommendations for closing the gap between INDCs and inter-provincial emission trading scheme (ETS).

To this end, the rest of this paper is structured as follows. In Section 6.2, the methodology and data for our analysis are introduced. Section 6.3 presents the empirical results derived from the ZSG-DEA model established in the paper. Section 6.4 compares the results with the literature and offers insights into policy implications. Conclusions are drawn in Section 6.5.

6.2 Methods and data

The schematic of the CEA allocation employed in this paper is illustrated in Figure 6.1. First, a suite of allocation principles and indicators are selected as output variables for CEA allocation that are expected to capture the three pillars of

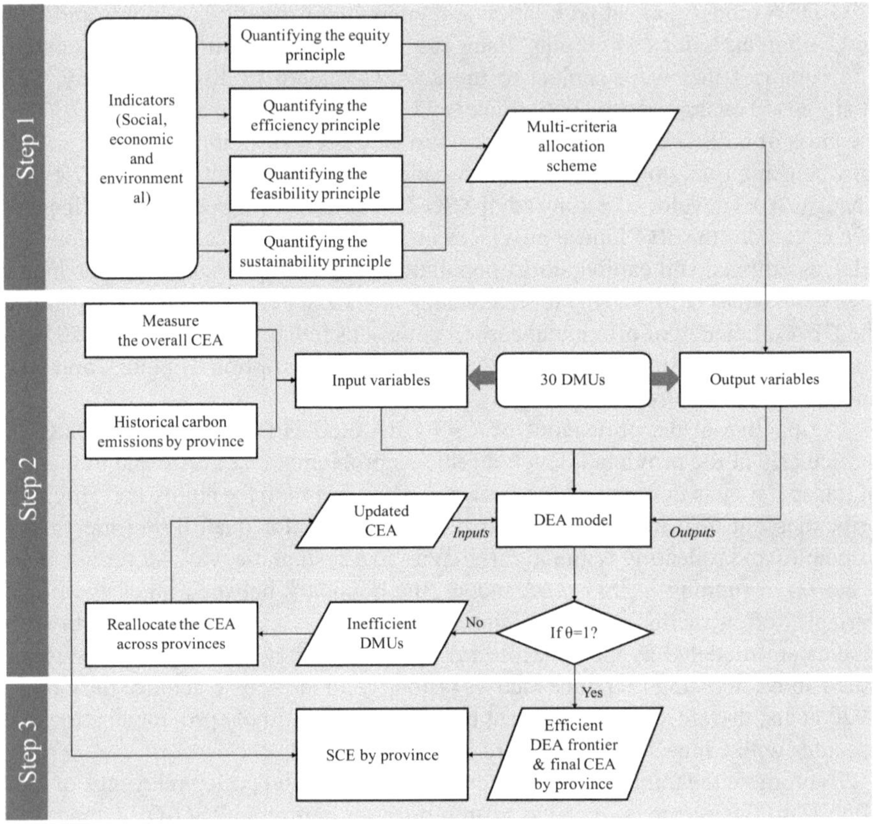

Figure 6.1 Schematic of CEA allocation

sustainability (i.e., social, economic, and environmental) while reflecting the four principles (namely equity, efficiency, feasibility, and sustainability). Second, historical carbon emissions by province lead to an estimate of initial CEA that would be reallocated across provinces given their DEA efficiency. Finally, we continue the process of reallocation until that all the provinces have reached the DEA efficiency of 100%. Comparing the final CEA with the current carbon emissions (CCE) of provinces, the space for carbon emissions (SCE) by province can be measured.

6.2.1 Multicriteria allocation scheme

As opposed to any single-criterion CEA allocation scheme, multicriteria ones accommodate indicators in a more systematic way reflecting common but differentiated responsibility for climate change mitigation. As suggested by Fang et al. (2018a), the equity, efficiency, feasibility, and sustainability principles are adopted in this paper for CEA allocation. Furthermore, on the basis of a comprehensive review of the literature, a suite of indicators intended for capturing these four

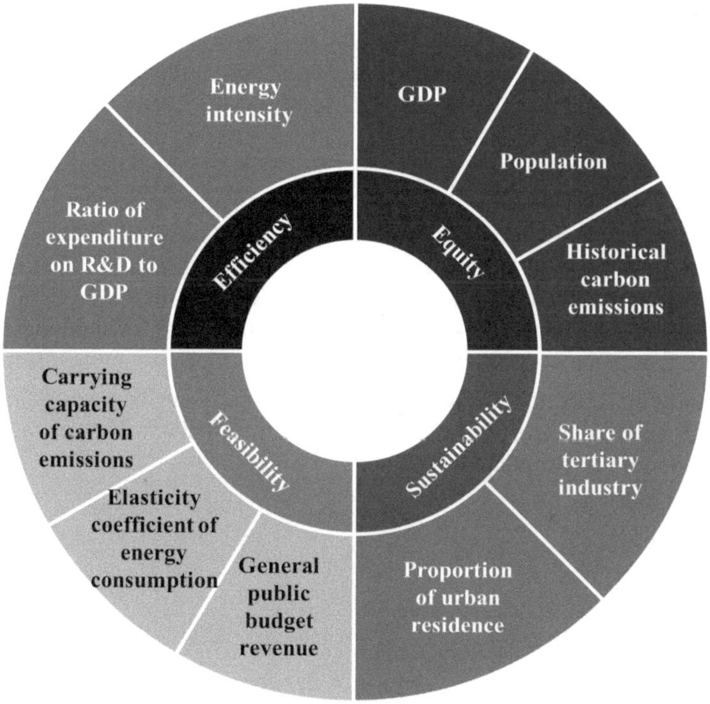

Figure 6.2 Multicriteria indicators for CEA allocation

Note: "Soc," "Eco," and "Env" represent social, economic, and environmental, respectively.

principles are selected in accordance with the social, economic and environmental pillars (Figure 6.2). However, some indicators, such as ecological resilience, are excluded due to the difficulty in measurement. The correlation analysis allows for further selection of the indicators, by which those correlated insignificantly with historical carbon emissions are deleted as well. Finally, the equity principle is expressed by population (social), GDP (economic), and historical carbon emissions (environmental), the efficiency principle by the ratio of expenditure on R&D to GDP (social) and energy intensity (economic), the feasibility principle by general public budget revenue (social), elasticity coefficient of energy consumption (economic) and the carrying capacity of carbon emissions (environmental), and the sustainability principle by the proportion of urban residence (social) and the share of tertiary industry (economic).

6.2.2 ZSG-DEA model

As noted, the hypothesis of the classical DEA that all DMUs are independent and have no impacts on others' actions is untenable for a competitive market, particularly given the fact that the overall CEA is bound to a value. What's more, through the classical DEA model one could investigate the DEA efficiency of each DMU, yet it is unable to bring them together into the DEA frontier for CEA reallocation. To maximize the DEA efficiency while making constant the overall CEA by 2030 as inputs, this paper develops an input-oriented ZSG-DEA model.

One prominent merit of this model is that it allows for the optimization of all the DMUs in order to reach the DEA frontier without altering the overall CEA. By improving the DEA efficiency of given DMUs and bringing them together to the DEA frontier, the inputs of inefficient DMUs (i.e., provincial CEA) must be declined. Meanwhile, the redundant CEA should be reallocated to other DMUs to keep the overall CEA unchanged. For instance, provided that DMU_k ($k=1,2,...,30$) with a CEA of X_k is an inefficient unit with its DEA efficiency θ, the CEA of DMU_k has to reduce by $X_k \cdot (1-\theta)$ to reach the DEA frontier and other DMUs would increase their CEA according to the weights expressed as $\frac{X_k \cdot (1-\theta)}{\sum_{i=1,i\neq k}^{30} X_i} \cdot X_i$. In principle, the inefficient DMUs continue to be reallocated until all reach the DEA frontier. After several rounds of reallocation, the DEA efficiency of all the DMUs will be equal to 1. The ZSG-DEA model is formulated as follows:

$$E_{ZSG} \doteq \min \theta \tag{6.1}$$

$$s.t.\begin{cases} \sum_{i=1}^{30} \lambda_i \cdot y_i^{equity} \geq y_k^{equity} \\ \sum_{i=1}^{30} \lambda_i \cdot y_i^{efficiency} \geq y_k^{efficiency} \\ \sum_{i=1}^{30} \lambda_i \cdot y_i^{feasibility} \geq y_k^{feasibility} \\ \sum_{i=1}^{30} \lambda_i \cdot y_i^{sustainability} \geq y_k^{sustainability} \\ \sum_{i=1}^{30} \lambda_i \cdot X_i \left[1 + \frac{X_k \cdot (1-\theta)}{\sum_{i=1,i\neq k}^{30} X_i} \right] \leq \theta \cdot X_k \\ \sum_{i=1}^{30} \lambda_i = 1 \\ \lambda_i \geq 0, i = 1,2,...,30 \end{cases} \tag{6.2}$$

where θ refers to the DEA efficiency of DMUs; E_{ZSG} refers to the minimum DEA efficiency of DMU_k; λ_i refers to the weight of DMU of i; X_i refers to the initial CEA as input; y_i^{equity}, $y_i^{efficiency}$, $y_i^{feasibility}$ and $y_i^{sustainability}$ refer to the quantified principles of equity, efficiency, feasibility, and sustainability, respectively.

To realize the allocation of CEA, the overall CEA must be measurable. Assuming that the annual GDP growth is at a rate of 6.0% on average during the phase of "New Normal," and that there will be a linear decline in emission intensity, the overall CEA for the next 15 years could be calculated as:

$$TCA_t = GDP_t \cdot (1+p)^{t-2015} \cdot I_{2015} \cdot (1-q)^{t-2015} \tag{6.3}$$

$$q = 1 - \sqrt[15]{\frac{I_{2030}}{I_{2015}}} = 1 - \sqrt[15]{\frac{I_{2005} \cdot (1-\alpha)}{I_{2015}}} \tag{6.4}$$

where TCA_t refers to the overall CEA in year t; GDP_t refers to the GDP in year t; p refers to the projected rate of GDP growth in next 15 years; I_{2015} refers to the emission intensity in 2015; q refers to the rate of decline in emission intensity in the next 15 years; α refers to the targeted reduction in emission intensity in 2030 compared with the level of 2005.

The initial CEA by province can be determined in accordance with the weights of historical carbon emissions for different provinces, with the aim of not causing a lot of derivation from a grandfathering perspective. As Formula (5) shows, $ICEA_p$ refers to the initial CEA of Province p and CE_p refers to the historical carbon emissions of Province p. The resulting $ICEA$ is adopted as input variables that would be reallocated through the ZSG-DEA model afterward.

$$ICEA_p = TCA \cdot \frac{CE_p}{\sum CE_p} \tag{6.5}$$

As illustrated in Figure 6.3, a multicriteria allocation scheme has been constructed as a benchmark for CEA reallocation. To quantify all the selected indicators, GM (1,1) is employed to project the trends between 2016 and 2030, which is a grey model proposed by Deng (1982) and has been widely applied to the projections of resources and emissions (Hamzacebi & Es, 2014; Pao et al., (2012). Given the complexity that multiple indicators may show similar trends along the time-series, we run a principal component analysis to aggregate these indicators for each principle, and the resulting components within single principles are weighted equally. Hence, each principle can be quantified through the projections of all the indicators involved. The equations are as follows:

$$PR = \sum w \cdot F_i = w \cdot \sum \sum \alpha_{ij} \cdot ind_{ij} \tag{6.6}$$

$$PR^* = \frac{PR - PR_{\min}}{PR_{\max} - PR_{\min}} \tag{6.7}$$

$$PR^* = \frac{PR_{\max} - PR}{PR_{\max} - PR_{\min}} \tag{6.8}$$

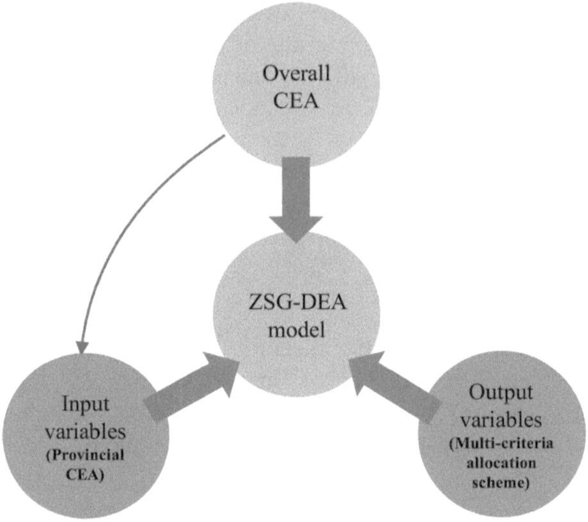

Figure 6.3 Schematic of the ZSG-DEA model developed in this paper

where *PR* and PR^* refer to the quantified principles before and after standardization, respectively; *w* refers to the weights of each prominent component *n*, *w=1/n*; a_{ij} refers to the score of each indicator *j* within the resulting components of the principle *i*; ind_{ij} refers to indicator *j* for principle *i*. Unlike the equity and efficiency principles, a higher ranking of the feasibility principle would lead to less share of CEA according to the principal component analysis, because it is dominated by general public budget revenue that corresponds to enhanced ability to control carbon emissions and, therefore, to less CEA. The same applies to the sustainability principles. For this reason, the standardization of these two principles could be conducted with Formula (8).

6.2.3 Data

Table 6.1 summarizes the indicators adopted for the measurement of the overall CEA, the initialization of provincial CEA and the construction of the multicriteria allocation scheme. It is worth mentioning that the carrying capacity of CO_2 sequestration within the principle of feasibility is calculated according to the respective capacity of different land use types within a region to store CO_2 through photosynthesis. For details of the data and calculation, please see Fang et al. (2012, 2014, 2018b).

Besides, we utilize the Data Processing System (DPS) software to predict the aforementioned indicators for the period 2016–2030 with the GM (1,1) model. SPSS 16.0 is applied to investigate the correlation among indicators as well as the principal component analysis to quantify the principles. DEAP software and Excel are used for DEA efficiency calculation and provincial CEA reallocation.

Table 6.1 Indicators of the DEA-ZSG model and the data sources

Composition	*Indicators*	*Data sources*
The overall CEA	GDP of China	China Statistic Yearbooks 2006–2016
	Emission intensity of China	China Emission Accounts & Datasets (CEADs)
Input variables	Historical carbon emissions by province	CEADs
Output variables	GDP by province	Statistical Yearbooks of each province 2006–2016
	Population by province	
	Energy intensity by province	
	Ratio of expenditure on R&D to GDP by province	
	General public budget revenue by province	
	Elasticity coefficient of energy consumption by province	
	Proportion of urban residence by province	
	Share of tertiary industry by province	
	Carrying capacity of CO_2 sequestration by province	Fang et al. (2012, 2018b).

6.3 Results

6.3.1 Changes in carbon emission allowance through reallocation

Figure 6.4 tracks the DEA efficiency of the initial allocation and its reallocation. Ten out of the 30 provinces have DEA efficiency that is higher than the average of initial efficiency (0.44). Beijing, Hainan and Qinghai already reach the DEA frontier where the DEA efficiency is equal to 1, whereas Shanxi reports the lowest DEA efficiency of 0.08. The average DEA efficiency increases to 0.61 after the first reallocation, even though no additional provinces reach the DEA frontier. The DEA efficiency of Shanxi almost triples while still ranking the last. The average DEA efficiency amounts to 0.81 and 0.96, respectively, as a result of the second and third reallocation, whereas the number of provinces with the DEA efficiency of 1 remains constant. Tremendous changes are found in the fourth reallocation, in which Liaoning, Jilin, Jiangsu, Fujian, Jiangxi, Shandong, Henan, Guangdong, and Ningxia reach the DEA frontier in addition to Beijing, Hainan, and Qinghai. Nevertheless, the average DEA efficiency slightly increases from 0.96 to 0.99. Ultimately, all the provinces achieve the maximum DEA efficiency of 1 after the final reallocation.

Figure 6.5 depicts the changes in the provincial shares of the overall CEA. The average CEA of the 30 provinces granted in the initial allocation is estimated at 8.44 Gt with a standard variance of 6.20 Gt. Shanxi and Hainan are found to make up the largest and smallest share of the CEA, respectively, with an estimation of 26.62 Gt and 3.91 Gt. This is because of the fact that the initial allocation is solely dependent on the historical carbon emissions by province, which could be interpreted as

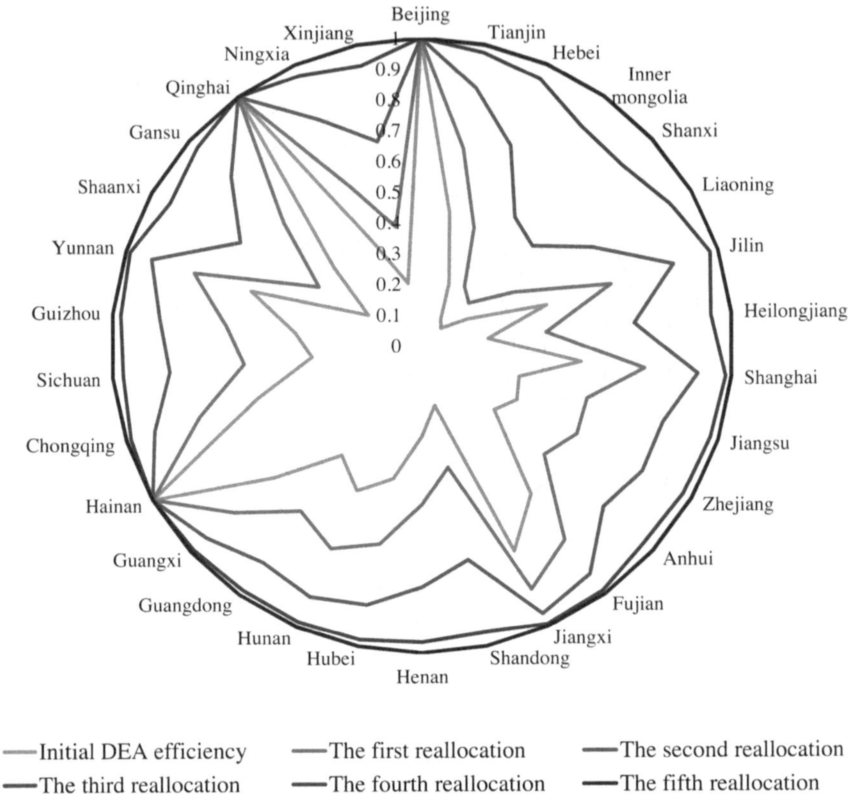

Figure 6.4 The DEA efficiency of initial allocation and reallocation

a grandfathering perspective. Through all the five rounds of reallocation, the CEA of Shanxi, Inner Mongolia, Shandong, Liaoning, Hebei, Shaanxi, and Xinjiang decreases by 165.6%, 34.7%, 43.6%, 28.4%, 35.3%, and 232.1%, respectively. Conversely, an increase in the share of CEA is witnessed in the remaining 21 provinces. For instance, Guangdong, Jiangxi, Beijing, Fujian, and Guangxi enlarge their CEA by 44.1%, 140.0%, 227.6%, 96.0%, and 109.8%, respectively.

By looking deeper into the geographical distribution of the changes to CEA by province, one may notice the agglomeration effect as evident from the fact that numerous provinces located in north China experience a significant decline in the CEA as compared to their initial shares and, conversely, many of the southern provinces have an increment of CEA. Slight changes occur mostly in middle China, where provinces either increase their CEA up to 4.00 Gt or decrease their CEA up to 3.00 Gt.

Figure 6.6 represents the distribution of frequency that refers to the number of provinces whose CEA falls into a range of values prior to and during the

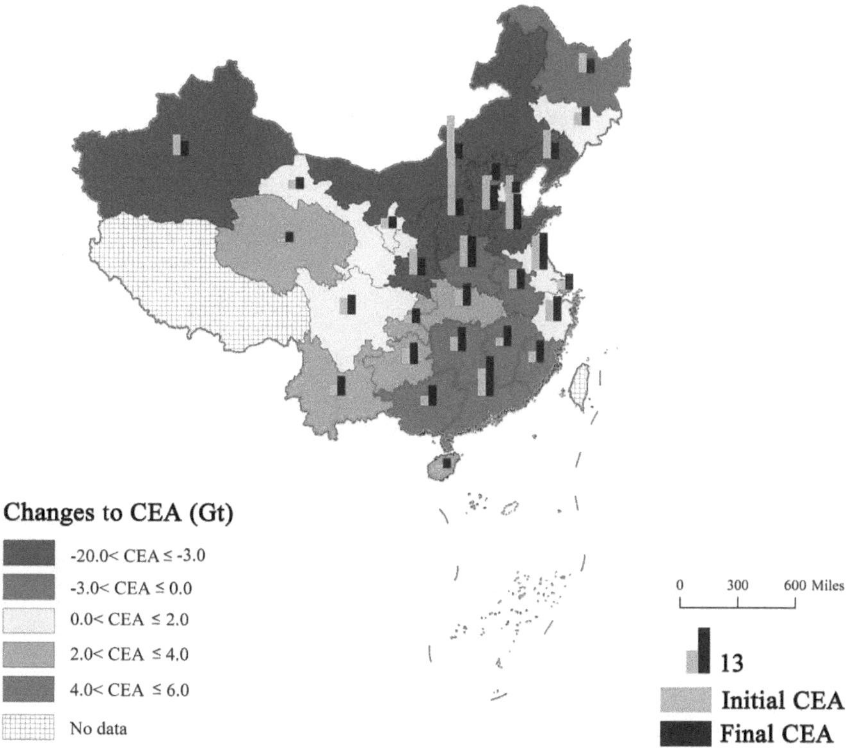

Figure 6.5 Geographical distribution of provincial CEA through reallocation

Note: All the provinces are marked on the basis of the changes to the provincial CEA between the initial allocation and final reallocation. Tibet, Taiwan, Hong Kong, and Macau are not displayed due to the lack of data.

reallocation. As shown, the distribution of the initial CEA covers the largest spectrum ranging from 1.29 Gt to 26.62 Gt with a standard variance of 6.20 Gt, suggesting that a CEA between 2.24 Gt and 14.64 Gt is entitled to more than two thirds of the provinces. After the first reallocation, the spectrum is shortened to between 2.23 Gt and 19.80 Gt, and two thirds of the provinces own a CEA ranging from 3.87 Gt to 13.02 Gt. Through the second reallocation, the CEA of 70% of the provinces stays within a range of 5.01–11.87 Gt. The third reallocation corresponds to the shortest spectrum of CEA shares ranging from 4.05 Gt to 16.47 Gt. The fourth and fifth reallocation finally witnesses that the CEA of each province remains stable (between 4.21 Gt and 16.76 Gt), while the standard variance decreases to 3.26 Gt eventually. As a result, one may conclude that this reallocation driven by the ZSG-DEA model could not only maximize the DEA efficiency of individual provinces but also contribute to the convergence of all the provinces by shrinking the CEA gap to some extent.

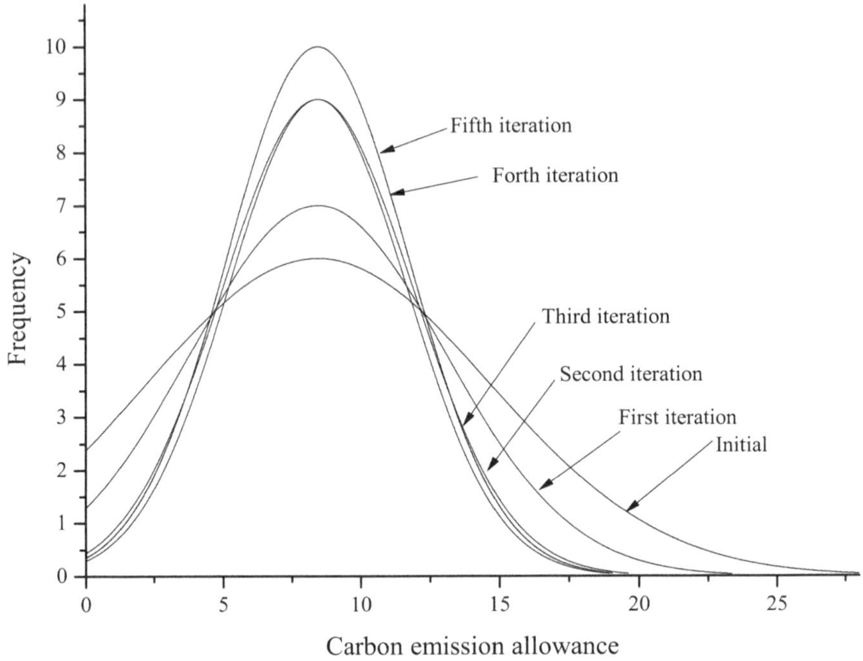

Figure 6.6 Frequency distribution of provincial CEA through reallocation

6.3.2 Final carbon emission allowance of provinces

Figure 6.7 delineates the ranking of provinces as per the principles of equity, efficiency, feasibility, and sustainability, respectively. As shown, the first two principles are gained prominence in eastern provinces such as Shandong, Guangdong, and Jiangsu. In the case of Guangdong, for instance, our principal component analysis indicates that GDP (89.2%) and population (9.9%) contribute the most to the equity principle, while the ratio of expenditure on R&D to GDP dominates the efficiency principle (over 90.0%). More than 20 provinces (e.g., Guizhou, Yunnan, Gansu) have the advantage of reflecting the principles of feasibility and sustainability through the CEA reallocation, many of which are located in southwest and northwest China. In the case of Guizhou, the general public budget revenue contributes the most (over 90.0%) to the feasibility principle, while the proportion of urban residents (49.1%) and the proportion of the tertiary industry (50.9%) contribute equally to the sustainability principle.

Figure 6.8 displays the discrepancy of geographical distribution between the annual average and per capita CEA of 30 provinces. The spatial pattern of the annual average CEA is prominent in the sense that provinces in south and east China tend to account for a large share of the overall CEA. This is particularly true for Guangdong, Shandong, Jiangsu, and Henan, whose CEA is at least over 0.70 Gt. On the contrary, those in the northeast and northwest China occupy a smaller CEA

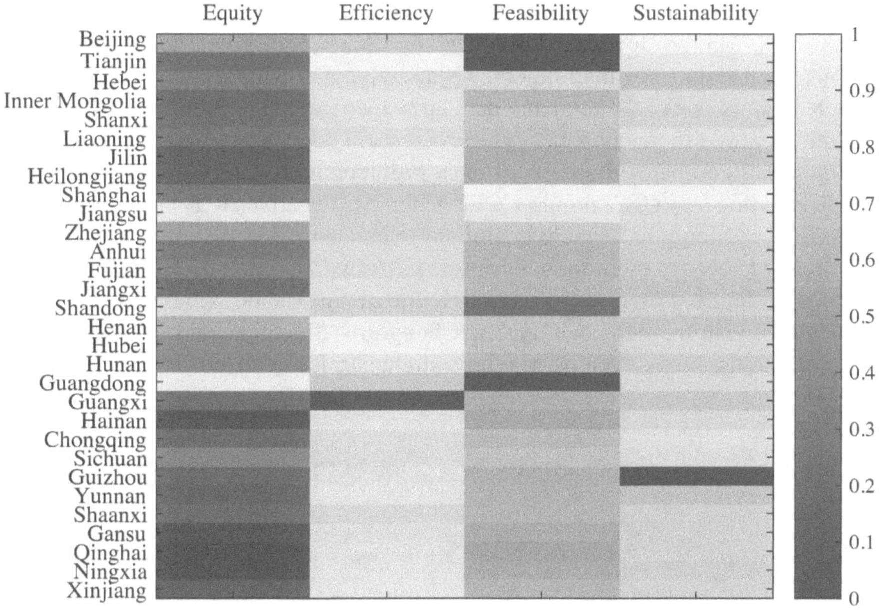

Figure 6.7 The ranking of all the provinces based on the four principles

Note: After the quantification of each principle by province, we standardize these 30 scores within each of the four principles, respectively, resulting in a range of values between 0–1. Hence the spectrum primarily shows the provincial ranking within each principle instead of quantifying the contributions of various principles to single provinces.

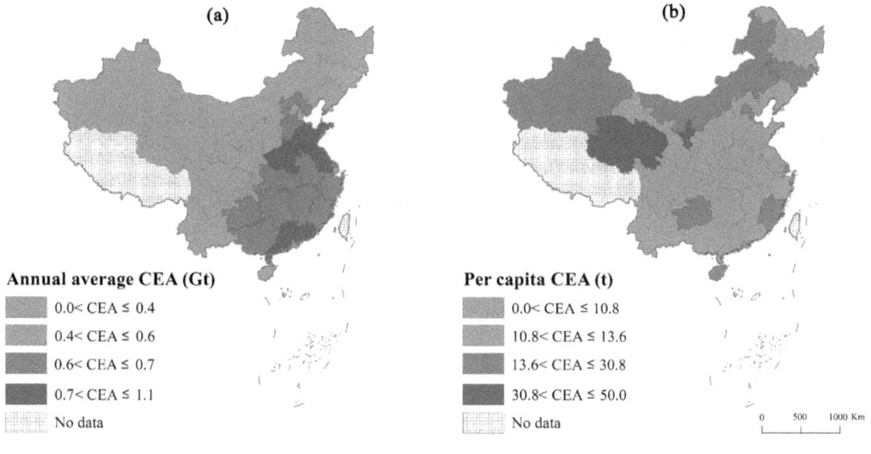

Figure 6.8 Geographical distribution of (a) the annual average CEA and (b) the per capita CEA by province

Note: Tibet, Taiwan, Hong Kong, and Macau are not displayed due to the lack of data.

(less than 0.60 Gt). A definitely different pattern occurs when it comes to the per capita CEA. The northwestern provinces of China (e.g., Ningxia, Qinghai) own a larger per capita CEA than others. These two patterns underline the need for a complementary use of the total and per capita metrics in interpreting the allocation scheme proposed in this paper.

Figure 6.9 compares the CEA of each province with their CCE and examines the SCE defined as CEA minus CCE. Amongst, negative SCE can be found in Shanxi, Inner Mongolia, Shaanxi, Xinjiang, Shandong, and Liaoning, whose CCE has already exceeded their annual average CEA by 0.99 Gt, 0.44 Gt, 0.19 Gt, 0.13 Gt, 0.05 Gt and 0.04 Gt, respectively. Heilongjiang, Hebei, Ningxia, Gansu, Anhui, Hainan, Tianjin, Sichuan, Qinghai, Chongqing, Shanghai, and Guizhou are approaching to the break-even point with a slight surplus of SCE (less than 0.30 Gt).

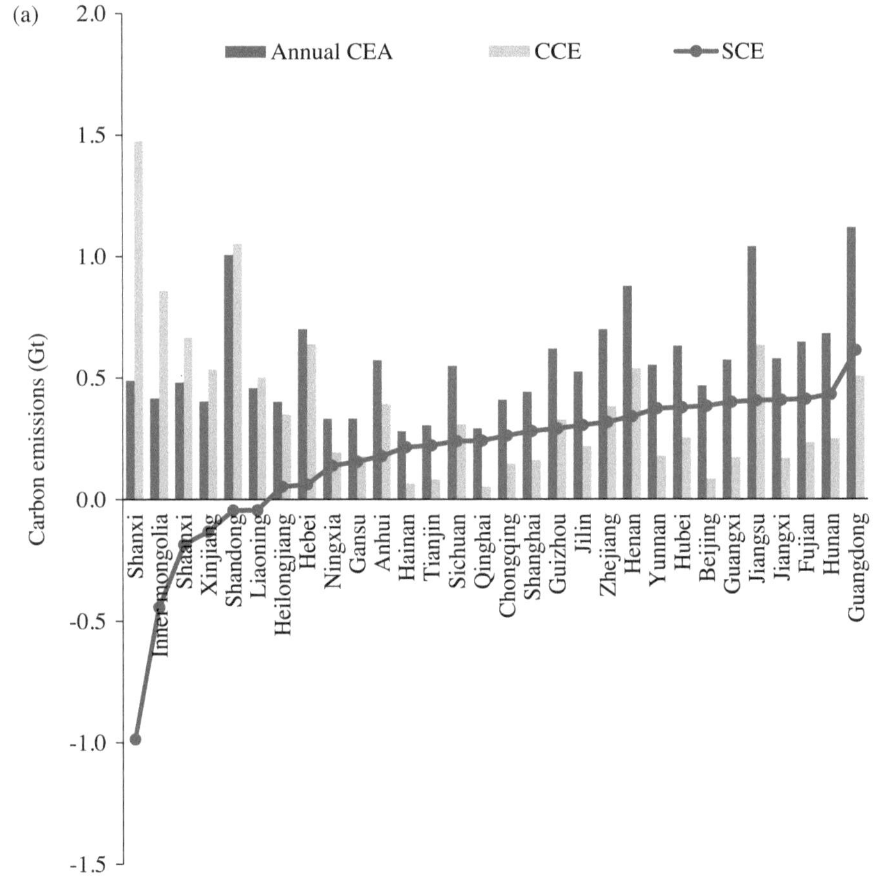

Figure 6.9 Annual carbon emission balance by province (*Continued*)

Note: Negative SCE denotes that annual average CEA could not offset CCE. SCE break-even point denotes that annual average CEA exceeds CCE by less than 0.30 Gt. Positive SCE denotes that annual average CEA exceeds CCE by at least 0.30 Gt.

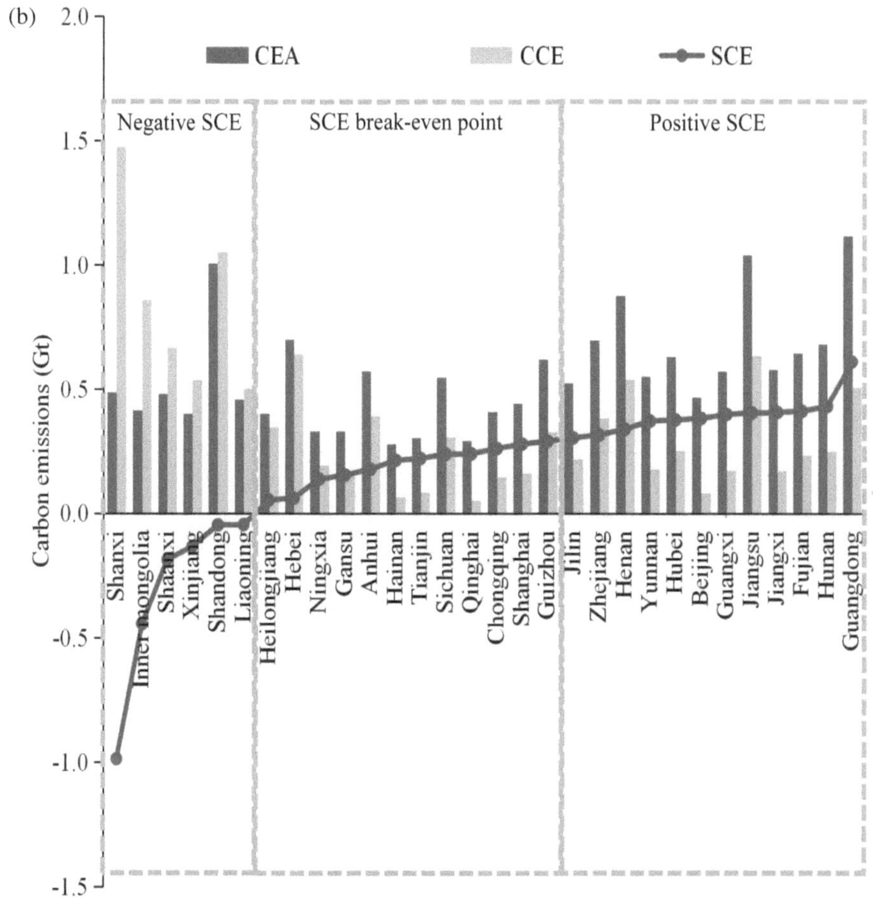

Figure 6.9 (Continued)

On the contrary, Guangdong, Hunan, Fujian, Jiangxi, Guangxi, Beijing, Hubei, Yunnan, Henan, Zhejiang, and Jilin are under relatively low levels of stress to control carbon emissions. However, this does not necessarily represent an absolute SCE, because the carbon emissions of most provinces have not peaked yet and thus will continue to rise (Li et al., 2018a; Zhang et al., 2018).

6.4 Discussion

6.4.1 Comparison with literature

In general, this paper proposes a moderate CEA allocation scheme, as witnessed by Figure 6.10 that provides a comparison between existing studies on CEA allocation across 30 Chinese provinces with ours. In some research, however, the highest provincial CEA exceeds the lowest one by over 26 times! That can hardly

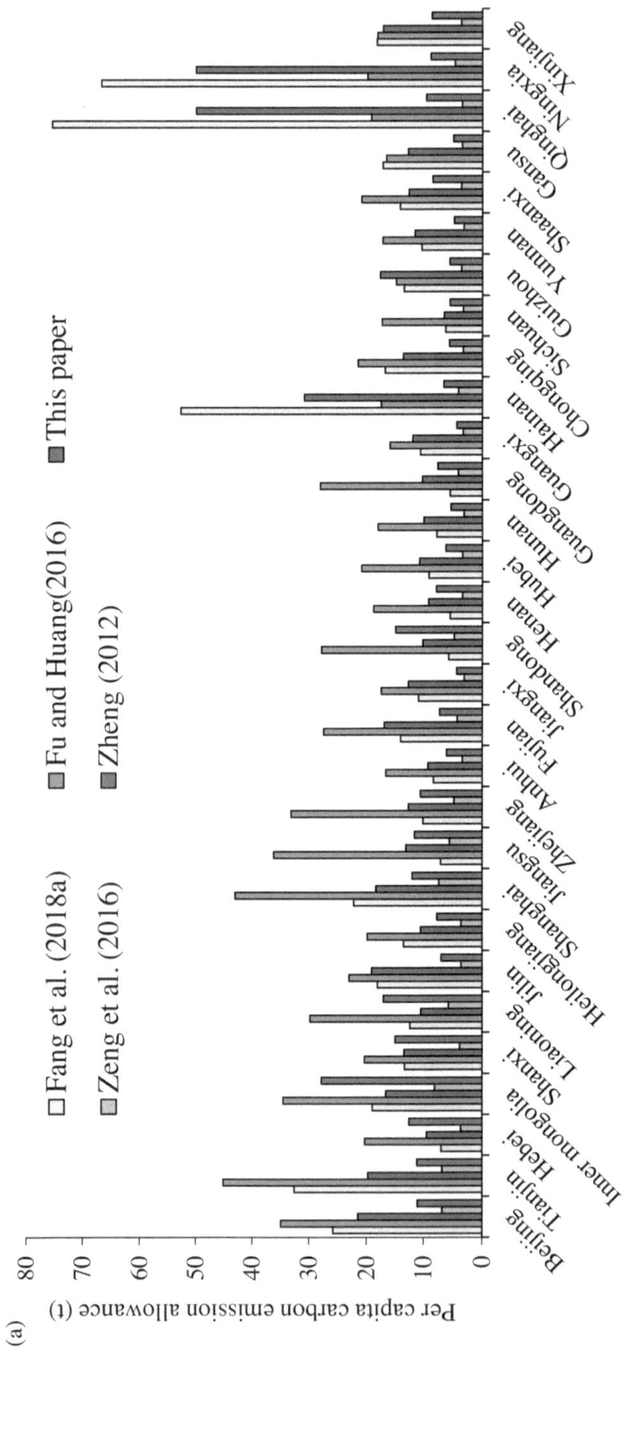

Figure 6.10 Comparison of different schemes for the allocation of China's carbon emission allowance derived from literature (*Continued*)

Note: All the results have been converted into (a) annual average CEA and (b) per capita CEA, respectively.

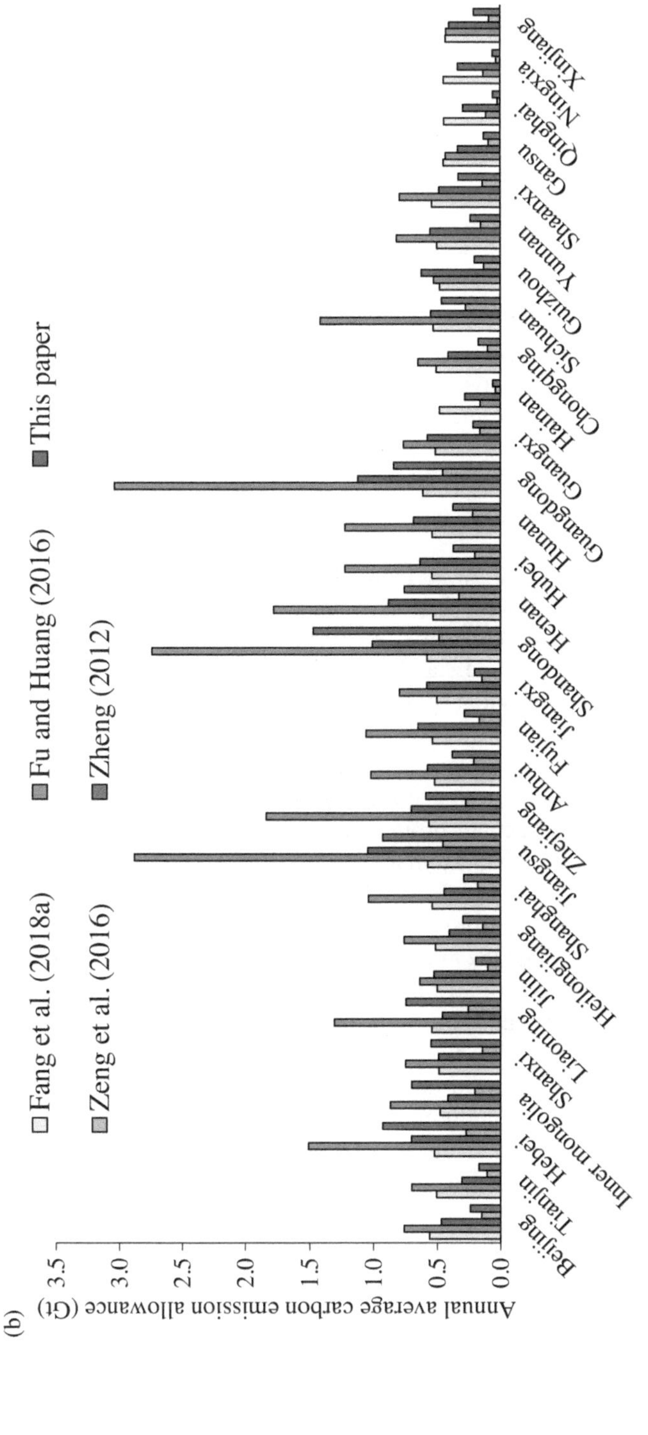

Figure 6.10 (Continued)

be implemented because of the absence of the equity principle in any recognizable sense. Except for Fang et al. (2018a), all these papers set the overall CEA in compliance with China's 2020 goal and conducted allocation by establishing ZSG-DEA models that took into account only a few indicators (e.g., GDP per capita, energy consumption, and/or population) as either input or output variables without unambiguous criteria. Although a number of indicators were chosen by Fang et al. (2018a) and the overall CEA was determined by China's INDCs toward 2030, their allocation procedure corresponds to a composite indicator approach that incorporates multiple indicators into a single-score metric for ease of understanding (Zhou and Wang, 2016). While this approach has received much attention from academia, it differs in nature from DEA method which is capable of integrating a variety of principles and indicators in a more sophisticated manner. For this reason, our improved ZSG-DEA model allows for a more equitable, efficient, feasible, and sustainable allocation scheme that gives preference to those of the provinces that would have considerable potential to increase the outputs without causing increase in the overall CEA of China.

6.4.2 Policy implications

Our study offers novel insights into the implementation of China's INDCs at the provincial level. The allocation scheme proposed in this paper allows policymakers to determine not only the CEA of Chinese provinces but also the corresponding SCE, which is a key to understanding differentiated emission reduction commitments and pressures confronted by individual provinces. The findings are particularly suited for pinpointing provinces with negative SCE (e.g., Shanxi, Inner Mongolia, Shaanxi)—those that face the challenge of achieving an absolute decoupling of their CO_2 emissions from economic growth. Furthermore, it should be noted that our analysis provides an optimistic estimate of provinces running a slight surplus of SCE (e.g., Anhui, Hainan, Hebei), since the SCE is defined as CEA minus CCE and overlooks the fact that the emissions of many provinces show negative or weak decoupling from economic growth and thus will continue to rise (Zhao and Li, 2018). Though the remaining provinces (e.g., Yunnan, Jiangxi, Guangxi) are not necessarily involve absolute emission reduction targets, there is still a need for controlling their emissions and reinforcing decoupling from economic growth.

In addition to laying a basis for formulating regional strategies for climate change mitigation and emission reduction, the findings would also be informative for the market construction of Chinese ETS—an inter-provincial cap-and-trade system that was formally launched in the end of 2017 and remains limited to electric power generation industry that accounts merely for a fraction of carbon emissions within Chinese territory. In other words, the proposed allocation scheme would help close the gap between INDCs and ETS, both of which the Chinese government is committed to. Nevertheless, the determination of caps for trading should be made cautiously. Only a tiny fraction of CEA of provinces with positive SCE should be allowed for sale. Those provinces with negative SCE must be prevented from selling CEA but they can be a buyer. In principle, provinces might sell their redundant CEA proportional to the positive SCE.

To our knowledge, this study, for the first time, takes into consideration the difference of provincial performance on the social, economic, and environmental pillars of sustainability simultaneously, by bringing together a systematic set of diverse principles and indicators into the ZSG-DEA model. Because of this, the resulting allocation scheme is able not only to reconcile the INDCs with regional development goals but to contribute to the trade-offs between various provinces—the key to consensus building on implementable schemes for regional CEA allocation. Moreover, the underlying methodology is also appropriate for use in allocating the permits of additional emissions or resources at multiple scales in the pursuit of optimal system efficiency.

6.4.3 Uncertainties and limitations

We acknowledge that our analysis remains the following limitations that should be addressed in future research: (1) the accounting of historical carbon emissions as a basis for initial CEA merely considers emissions from fossil fuels, and therefore neglects those from cement production and land use, especially for those embodied in interregional trade of goods and services (Liang, et al., 2017; Shao et al., 2018). This leads to call for input–output analysis (IOA), by which the consumption-based accounting of provincial carbon emissions could enhance the scientific robustness of initial allocation of CEA (Mi et al., 2016); (2) when estimating China's overall CEA by 2030, we presume a linear reduction of emission intensity and a fixed rate of GDP growth, which is not consistent with the reality and needs to be improved through the development of scenario analysis aiming to track the dynamics of emission intensity and GDP growth but also other input and output variables adopted in the modeling where economic sense is insufficiently considered in its current form; (3) in addition to the goal of reducing emission intensity by 60–65% in 2030, China's INDCs are also committed to peaking the carbon emissions around 2030. To fulfill these dual goals, it makes sense to conduct additional assignments of CEA for each province on a yearly basis in accordance to the anticipated trajectories of provincial carbon emissions; (4) DEA or ZSG-DEA method has the merit of assessing the technical efficiency among DMUs, whereas this may bring the risk of overemphasizing technical efficiency and overlooking other principles; (5) CO_2 emissions are often found to be accompanied by other pollutants (e.g., NO_x, SO_2), even though in this paper we take the former as the only undesirable emissions to control. With the consideration of multi-pollutants, there seems to be much room for an investigation into the trade-offs and co-benefits of regulations for different emissions when doing allocation (Cai et al.,2016; Yang et al.,2018); and (6) the principles of equity, efficiency, feasibility, and sustainability are treated without weighting, but no two entities actually hold the same importance. Therefore, it may make sense to employ unequal weighting that reflects stated or revealed preferences and judgments of provincial performance on social, economic, and environmental sustainability.

6.5 Conclusions

This paper begins with the measurement of the overall CEA by 2030 and further establishes an improved ZSG-DEA model to allocate the CEA to the 30 Chinese provinces, where the initial provincial CEA determined by the historical carbon emissions is treated as inputs and a suite of indicators pertaining to the equity, efficiency, feasibility, and sustainability principles are selected as outputs. Ultimately, all the provinces reach the DEA frontier with a CEA ranging from 4.21 Gt to 16.77 Gt. Specifically, some northern provinces are found to cut down their CEA while most southern provinces experience an increase in their CEA, as opposed to the initial shares. Comparing the final share of CEA by province with CCE, we observe that positive SCE is witnessed in 12 provinces including Guangdong, Hunan, Fujian, etc. The SCE of remaining provinces such as Heilongjiang, Hebei, Ningxia is approaching to the break-even point. The findings are particularly suited for pinpointing provinces with negative SCE, highlighting the need for region-specific policies and inter-provincial collaboration that would help China achieve a sustainable transition. As such, the multicriteria allocation scheme proposed in this paper not only lays a scientific basis for the Chinese government to make the INDCs implementable at the provincial level but also serves as a reference for fueling further scientific discussions and development of schemes for the allocation of responsibility at multiple scales within and beyond China. Moreover, the analysis is of benefit to closing the gap between INDCs and ETS, with the recognition that a tiny fraction of CEA of provinces with positive SCE can be allowed for sale, and that those with negative SCE can be a buyer merely.

However, the complexity and heterogeneity embodied in the regional allocation of CEA convince us that no single scheme should be interpreted as the "golden standard." Our allocation scheme, which takes into consideration the difference of provincial performance on the social, economic, and environmental pillars of sustainability, is not an exception. Thus, it does not pretend to be useful for all purposes, and it is not surprising that more elaborated or alternative schemes for the allocation of China's CEA at multiple scales will continue to emerge as the next step toward a more comprehensive implementation of China's INDCs. To that end, we suggest that further improvements need to focus on the following prioritized directions: (1) methodological integration of DEA and IOA to fully capture the provincial responsibility for carbon emissions from a consumption perspective; (2) development of scenario analysis to track the dynamics of input and output variables (e.g., emission intensity, GDP) so as to ensure accurate estimates in future trends; (3) inclusion of simulation of provincial peak emissions with the aim of achieving the dual control of emission intensity and total amount of CO_2 by 2030, as regulated in China's INDCs; and (4) linking the resultant CEA to the national ETS for policy simulation in marketing.

References

Banker, R. D., Charnes, A., & Cooper, W. W. (1984). Some models for estimating technical and scale inefficiencies in data envelopment analysis. *Management Science*, *30*(9), 1078–1092.

Cai, B., Bo, X., Zhang, L., Boyce, J. K., Zhang, Y., & Lei, Y. (2016). Gearing carbon trading towards environmental co-benefits in China: Measurement model and policy implications. *Global Environmental Change*, *39*, 275–284.

Charnes, A., Cooper, W. W., & Rhodes, E. (1978). Measuring the efficiency of decision making units. *European Journal of Operational Research*, *2*(6), 429–444.

Chiu, Y. H., Lin, J. C., Hsu, C. C., & Lee, J. W. (2013). Carbon emission allowances of efficiency analysis: Application of super SBM ZSG-DEA model. *Polish Journal of Environmental Studies*, *22*(3), 653–666.

Chiu, Y. H., Lin, J. C., Su, W. N., & Liu, J. K. (2015). An efficiency evaluation of the EU's allocation of carbon emission allowances. *Energy Sources, Part B: Economics, Planning, and Policy*, *10*(2), 192–200.

Chu, K. W. (2017). Screening environmental challenges in China: Three modes of ecocinema. *Journal of Chinese Governance*, *2*(4), 437–459.

Deng, J. L. (1982). Control problems of grey systems. *System & Control Letters*, *1*(5), 288–294.

Energy Information Administration (EIA) (2009). *International energy outlook 2009*. Energy Information Administration.

Fang, K., Dong, D., Lin, Z., & Shen, W. (2012). Calculation method of energy ecological footprint based on global net primary productivity. *Acta Ecologica Sinica*, *32*, 2900–2909.

Fang, K., Heijungs, R., & de Snoo, G. R. (2014). Theoretical exploration for the combination of the ecological, energy, carbon, and water footprints: Overview of a footprint family. *Ecological Indicators*, *36*, 508–518.

Fang, K., Zhang, Q., Ye, R., & Zhou, Y. (2018a). Allocating China's carbon emission allowance to the provincial quotas in the context of the Paris agreement. *Acta Scientiae Circumstantiae*, *38*, 1224–1234.

Fang, K., Zhang, Q., Yu, H., Wang, Y., Dong, L., & Shi, L. (2018b). Sustainability of the use of natural capital in a city: Measuring the size and depth of urban ecological and water footprints. *Science of the Total Environment*, *631*, 476–484.

Gomes, E. G., & Lins, M. E. (2008). Modelling undesirable outputs with zero sum gains data envelopment analysis models. *Journal of the Operational Research Society*, *59*(5), 616–623.

Grasso, M., & Roberts, J. T. (2014). A compromise to break the climate impasse. *Nature Climate Change*, *4*(7), 543–549.

Green, F., & Stern, N., (2015). China's "New Normal": Structural Change, Better Growth, and Peak Emissions. Policy Brief.

Guan, T., & Delman, J. (2017). Energy policy design and China's local climate governance: Energy efficiency and renewable Energy policies in Hangzhou. *Journal of Chinese Governance*, *2*(1), 68–90.

Hamzacebi, C., & Es, H. A. (2014). Forecasting the annual electricity consumption of Turkey using an optimized grey model. *Energy*, *70*, 165–171.

Han, R., Tang, B. J., Fan, J. L., Liu, L. C., & Wei, Y. M. (2016). Integrated weighting approach to carbon emission quotas: An application case of Beijing-Tianjin-Hebei region. *Journal of Cleaner Production*, *131*, 448–459.

Hof, A. F., den Elzen, M. G., Admiraal, A., Roelfsema, M., Gernaat, D. E., & van Vuuren, D. P. (2017). Global and regional abatement costs of nationally determined contributions (NDCs) and of enhanced action to levels well below 2°C and 1.5°C. *Environmental Science & Policy*, *71*, 30–40.

Intergovernmental Panel on Climate Change (2015). *Climate change 2014: Mitigation of climate change (Vol. 3)*. Cambridge University Press.

Ju, Y., & Fujikawa, K. (2019). Modeling the cost transmission mechanism of the emission trading scheme in China. *Applied Energy*, *236*, 172–182.

Kriegler, E., O'Neill, B., Hallegatte, S., Kram, T., Moss, R. H., Lempert, R., & Wilbanks, T. J., (2010). Socioeconomic Scenario Development for Climate Change Analysis. CIRED Working Papers, 2010, 23.

Li, L., Li, Y., Ye, F., & Zhang, L. (2018b). Carbon dioxide emissions quotas allocation in the Pearl River delta region: Evidence from the maximum deviation method. *Journal of Cleaner Production*, *177*, 207–217.

Liang, S., Qu, S., Zhu, Z., Guan, D., & Xu, M. (2017). Income-based greenhouse gas emissions of nations. *Environmental Science & Technology*, *51*(1), 346.

Lin, J. Y. (2011). China and the global economy. *China Economic Journal*, *4*(1), 1–14.

Lins, M. P., Gomes, E., de Mello, E. G., & de Mello, J. C. C. S. (2003). Olympic ranking based on a zero sum gains DEA model. *European Journal of Operational Research*, *148*(2), 312–322.

Li, F., Xu, Z., & Ma, H. (2018a). Can China achieve its CO_2 emissions peak by 2030? *Ecological Indicators*, *84*, 337–344.

Marangoni, G., Tavoni, M., Bosetti, V., Borgonovo, E., Capros, P., Fricko, O., Gernaat, D. E. H. J., Guivarch, C., Havlik, P., Huppmann, D., Johnson, N., Karkatsoulis, P., Keppo, I., Krey, V., Broin, E., Price, J., & van Vuuren, D. P. (2017). Sensitivity of projected long-term CO_2 emissions across the shared socioeconomic pathways. *Nature Climate Change*, *7*(2), 113.

Mi, Z., Zhang, Y., Guan, D., Shan, Y., Liu, Z., Cong, R., Yuan, X., & Wei, C. (2016). Consumption-based emission accounting for Chinese cities. *Applied Energy*, *184*, 1073–1081.

Miao, Z., Geng, Y., & Sheng, J. (2016). Efficient allocation of CO_2 emissions in China: A zero sum gains data envelopment model. *Journal of Cleaner Production*, *112*, 4144–4150.

Pan, J. (2003). Emissions rights and their transferability: Equity concerns over climate change mitigation. *International Environmental Agreements*, *3*(1), 1–16.

Pang, R. Z., Deng, Z. Q., & Chiu, Y. H. (2015). Pareto improvement through a reallocation of carbon emission quotas. *Renewable and Sustainable Energy Reviews*, *50*, 419–430.

Pan, X., Teng, F., & Wang, G. (2014). Sharing emission space at an equitable basis: Allocation scheme based on the equal cumulative emission per capita principle. *Applied Energy*, *113*, 1810–1818.

Pao, H. T., Fu, H. C., & Tseng, C. L. (2012). Forecasting of CO_2 emissions, energy consumption and economic growth in China using an improved grey model. *Energy*, *40*(1), 400–409.

Raupach, M. R., Davis, S. J., Peters, G. P., Andrew, R. M., Canadell, J. G., Ciais, P., Friedlingstein, P., Jotzo, F., van Vuuren, D., & Le Quere, C. (2014). Sharing a quota on cumulative carbon emissions. *Nature Climate Change*, *4*(10), 873–879.

Ringius, L., Torvanger, A., & Holtsmark, B. (1998). Can multi-criteria rules fairly distribute climate burdens?: OECD results from three burden sharing rules. *Energy Policy*, *26*(10), 777–793.

Rogelj, J., Den Elzen, M., Höhne, N., Fransen, T., Fekete, H., Winkler, H., Schaeffer, R., Sha, F., Riahi, K., & Meinshausen, M. (2016). Paris Agreement climate proposals need a boost to keep warming well below 2°C. *Nature*, *534*(7609), 631–639.

Shao, L., Li, Y., Feng, K., Meng, J., Shan, Y., & Guan, D. (2018). Carbon emission imbalances and the structural paths of Chinese regions. *Applied Energy*, *215*, 396–404.

Shao, S., Liu, J., Geng, Y., Miao, Z., & Yang, Y. (2016). Uncovering driving factors of carbon emissions from China's mining sector. *Applied Energy*, *166*, 220–238.

Tokarska, K. B., Gillett, N. P., Weaver, A. J., Arora, V. K., & Eby, M. (2016). The climate response to five trillion tonnes of carbon. *Nature Climate Change*, *6*(9), 851–855.
United Nations Framework Convention on Climate Change, (2010). UNFCCC. Outcome of the work of the Ad Hoc Working Group on long-term Cooperative Action under the Convention, Home, Cancun Agreements, Draft decision/CP.16.
Wang, J. N., Cai, B. F., Cao, D., Zhou, Y., & Liu, L. (2011). Scenario study on regional allocation of CO_2 emissions allowance in China. *Acta Scientiae Circumstantiae*, *31*(4), 680–685.
Wang, L., Chen, W., Zhang, H., & Ma, D. (2017). Dynamic equity carbon permit allocation scheme to limit global warming to two degrees. *Mitigation and Adaptation Strategies for Global Change*, *22*(4), 609–628.
Wang, K., Zhang, X., Wei, Y. M., & Yu, S. W. (2013). Regional allocation of CO_2 emissions allowance over provinces in China by 2020. *Energy Policy*, *54*, 214–229.
Wen, L., & Zhang, E. (2016). Regional allocation of carbon emissions in China based on zero sum gains data envelopment analysis model. *Environmental Engineering Research*, *21*(1), 91–98.
Yang, H., Liu, J., Jiang, K., Meng, J., Guan, D., Xu, Y., & Tao, S. (2018). Multi-objective analysis of the co-mitigation of CO_2 and $PM_{2.5}$ pollution by China's iron and steel industry. *Journal of Cleaner Production*, *185*, 331–341.
Ye, B., Jiang, J., Li, C., Miao, L., & Tang, J. (2017). Quantification and driving force analysis of provincial-level carbon emissions in China. *Applied Energy*, *198*, 223–238.
Yi, W. J., Zou, L. L., Guo, J., Wang, K., & Wei, Y. M. (2011). How can China reach its CO_2 intensity reduction targets by 2020? A regional allocation based on equity and development. *Energy Policy*, *39*(5), 2407–2415.
Yu, S. W., Wei, Y. M., & Wang, K. (2014). Provincial allocation of carbon emission reduction targets in China: An approach based on improved fuzzy cluster and Shapley value decomposition. *Energy Policy*, *66*, 630–644.
Zhang, Y. J., & Hao, J. F. (2017). Carbon emission quota allocation among China's industrial sectors based on the equity and efficiency principles. *Annals of Operations Research*, *255*(1–2), 117–140.
Zhang, Q., Liao, H., & Hao, Y. (2018). Does one path fit all? An empirical study on the relationship between energy consumption and economic development for individual Chinese provinces. *Energy*, *150*, 527–543.
Zhang, Y. J., Wang, A. D., & Da, Y. B. (2014). Regional allocation of carbon emission quotas in China: Evidence from the Shapley value method. *Energy Policy*, *74*, 454–464.
Zhao, M. M., & Li, R. (2018). Decoupling and decomposition analysis of carbon emissions from economic output in Chinese Guangdong Province: A sector perspective. *Energy & Environment*, *29*(4), 543–555.
Zhou, P., & Wang, M. (2016). Carbon dioxide emissions allocation: A review. *Ecological Economics*, *125*, 47–59.

7 Sectoral Coverage of Carbon Trading Scheme

7.1 More or less? Participants of emission trading scheme

Over 150 nations and regions have pledged their nationally determined contributions (NDCs) to limit the world-average temperature rise below 2°C and even 1.5°C relative to the pre-industrial levels (Glanemann et al., 2020; IPCC 2014; Weikmans et al., 2020). In attempting to avoid unaffordable emission abatement costs (EACs) as a result of increasingly stringent environmental regulations, market-based policies like emission trading scheme (ETS) that allows the emission allowances or reduction quotas trading has come into force in many countries and regions (Bertram et al., 2015; Fujimori et al., 2016; Liu et al., 2015a; Spash, 2010). For example, the European Union launched the ETS that serves as a powerful policy instrument in support of the climate targets (Skjærseth & Wettestad, 2008; Zhang & Wei, 2010). To date, the ETS has been implemented in a growing number of nations such as Switzerland, Australia, New Zealand, and South Korea, and plays a central role in meeting their climate targets and saving the EACs (Perdan & Azapagic, 2011).

As the world's largest greenhouse gas emitter, China has committed to reducing the carbon emissions per unit of GDP (i.e., carbon intensity) by over 65% in 2030 against the 2005 levels, reaching carbon peaking by 2030 and carbon neutrality by 2060 (Cui et al., 2021; Fang et al., 2021; Xie et al., 2020; Zhang et al., 2022; Zhou et al., 2021). To achieve these ambitious goals, China has devoted to the ETS pilots in the past decade (Dong et al., 2019; Heggelund et al., 2019; Liu et al., 2015b). In the recent years, China has sped up the pace toward the nation-wide ETS (Stoerk et al., 2019; Ying & Sovacool, 2021). However, despite these milestones, the limited scope of sectoral coverage remains as one major challenge facing China's ETS (Fan & Wang, 2014; Green, 2021; Mu et al., 2018; Qi & Cheng, 2018). In practice, China's ETS pilots vary considerably in their sectoral coverages. Of these, Hubei, Chongqing, etc., focus on traditional carbon-intensive sectors like power, cement, and steel production, whereas Beijing, Shanghai, and Shenzhen include the finance, public management, and commercial activities in their ETS (Jiang et al., 2016; Wang et al., 2017). Yet, the ongoing development of national ETS merely takes into account electricity production (Lin & Jia, 2019).

In addition to these learning-by-doing experiments, the theoretical underpinning of potential sectoral coverage has been explored (Fan & Wang, 2014; Karstensen

DOI: 10.4324/9781003433859-7

& Peters, 2018; Qi & Cheng, 2018). Generally speaking, the wider sectoral coverage that increases the variation of marginal abatement costs (MACs) across sectors can lead to greater potential cost savings of ETS, thereby improving the cost-effectiveness (Fan & Wang, 2014; Weitzman & Spence, 1978; Wu et al., 2022a). However, in practice, the extension of sectoral coverage will inevitably increase the administrative and transaction costs, and undermine the cost-effectiveness of ETS (Lin & Jia, 2020; Wang et al., 2017). In this sense, a partial sectoral coverage has been widely accepted in implementing the ETS despite the debates on the selection of sectors (Böhringer et al., 2014a; Fan & Wang, 2014; Springmann, 2012). Overall, the state of the art on this issue can be divided into two categories that concentrate on either the emission reduction potential or carbon leakage of sectors (Lu et al., 2020; Mu et al., 2018; Wang et al., 2018a; Wu et al., 2022b; Zhou et al., 2020). The rationale of the first category is naturally understandable, by which the carbon-intensive sectors will be the first included in the ETS given their high carbon emissions or high carbon intensity. This solution is deemed more practically sufficient to reduce the emissions and meet the regional climate targets even with partial sectoral coverage (Lamb et al., 2021, Qian et al., 2018; Wang et al., 2017).

The second category, originated from the spillover effect of unilateral environmental regulations, concerns about the impacts of sectoral coverage on the carbon leakage between sectors and regions (Baylis et al., 2013; Böhringer et al., 2014b; Jarke & Perino, 2017). This is particularly true for a partially-covered ETS, where the engaged sectors might shift their emissions to those who are not included in the ETS through trading, production reallocation or other channels, unintendedly leading to carbon leakage (Bistline & Rose, 2018; Di Maria & Van der Werf, 2008; Tan et al., 2018; Zhang et al., 2020). By assessing the trade intensity that is generally defined as the ratio of the exports and imports to the sum of domestic production and imports, some suggest that those sectors having less inter-sectoral trade in goods should be first included in the ETS in fighting carbon leakage (Juergens et al., 2013; Qian et al., 2018; Wang et al., 2017).

Even with the insightful views on sectoral coverage from current studies, they might disregard the fact that the ETS is primarily designed for minimizing the total EACs (Convery, 2009; Zhang & Wei, 2010). Despite the consensus that covering more sectors in the ETS can continuously reduce the EACs, the marginal cost savings (MCSs) that represent the marginal reduction of EACs due to the inclusion of an additional sector in the ETS can vary considerably because of the disparate MACs across sectors (De Cara & Jayet, 2011; Tang et al., 2020). In this sense, the estimation on varying MCSs of sectors would allow for a remarkable decline of EACs with a relatively small sectoral coverage. However, to our knowledge, the MCSs of sectors have not been sufficiently estimated. This absence actually undermines the cost-effective design of ETS with respect to sectoral coverage.

To fill the research gaps, the main contributions of this study are threefold as follows. First, rather than focusing on carbon- and trade-intensive sectors, we begin with looking at the variations in MCSs of each sector and minimize the EACs in any sectoral coverages, with the aim of assisting policymakers in developing a more cost-effective ETS. Second, we propose a costs-oriented approach for sectoral coverage (COASCO) for the MCSs quantification and conduct an empirical study that explores the impacts of sectoral coverage on EACs under China's NDCs by 2030. Third, we validate the Pareto principle that covering six sectors (e.g., *Electricity production*, *Metallurgy*, *Transport and storage*, *Petroleum and gas*, *Nonmetal mining*) out of 29 can reduce China's EACs by over 80% compared to covering *Electricity production* only. The results obtained in this paper could not only be informative for extending China's ETS but also serve as a robust reference for other economies dedicated to a more cost-effective ETS.

7.2 Methods and data

Rather than carrying on the debate over emission reduction potential versus carbon leakage, we propose determining the sectoral coverage through the COASCO method that selects and ranks the sectors according to their MCSs (Betz et al., 2010; Wang et al., 2018b). Unlike the general belief that sectors with lower MACs should be first covered in the ETS, Figure 7.1 provides a schematic representation of how to measure the MCSs of sectors and how to determine the sectoral coverage based on their varying MCSs. Specifically, suppose that Sectors *a*, *b*, and *c* have MAC curves, MAC_a, MAC_b, and MAC_c, where the historical carbon emission reduction (CER) starts at a_0, b_0, and c_0, respectively. As seen in Figure 7.1(a), compared to Sectors *b* and *c*, covering Sector *a* that has the lowest MACs would lead to fewer EACs to comply with the emission reduction quotas (ERQs) if only one sector is involved. Specifically, the EACs would be equal to the area of Polygons $a_0a_1p_{a1}p_{a0}$, $b_0b_1p_{b1}p_{b0}$, and $c_0c_1p_{c1}p_{c0}$, when the ETS includes Sectors *a*, *b*, and *c*, respectively. Provided that the ETS extends its coverage to two sectors, there are two options: Sectors *a* and *b*, or Sectors *a* and *c*. While Sector *a* increases its CER in both options, the marginal reduction of EACs (defined as MCSs) differs considerably between Sectors *b* and *c*. For example, when Sector *b* is additionally included in the ETS, the EACs would be reduced by $S_1+S_2+S_3$, namely, the MCSs of Sector *b* would be $S_1+S_2+S_3$ (Figure 7.1(b)). Although the EACs can also be reduced to some extent in the latter option, the MCSs of Sector *c* (which are estimated at S_1+S_4) are comparatively lower than those of Sector *b*. In this case, including Sector *b* is preferred even though the MACs of Sector *b* at b_0 are higher than those of Sector *c* at c_0. This simplified example illustrates how the ETS manages to minimize the total EACs; that is, the convergence of MACs of all sectors is achieved through ERQs trading. All this validates the methods used in this paper (i.e., the nonlinear programming) to simulate the ETS in terms of sectoral MACs at the macro scale. Meanwhile, it illustrates the way that we define and estimate the MCSs of each sector, and the reasons why sectors have varying MCSs.

Here we propose the COASCO method that conducts an iteration algorithm for estimating the MCSs of sectors and determines the rankings of ETS-covered sectors in order. Technically, we do four steps:

Step 1: Create new sets for sectoral coverage of ETS. Prior to estimating the MCSs of each sector for prioritization, we recursively create the new sets for sectoral

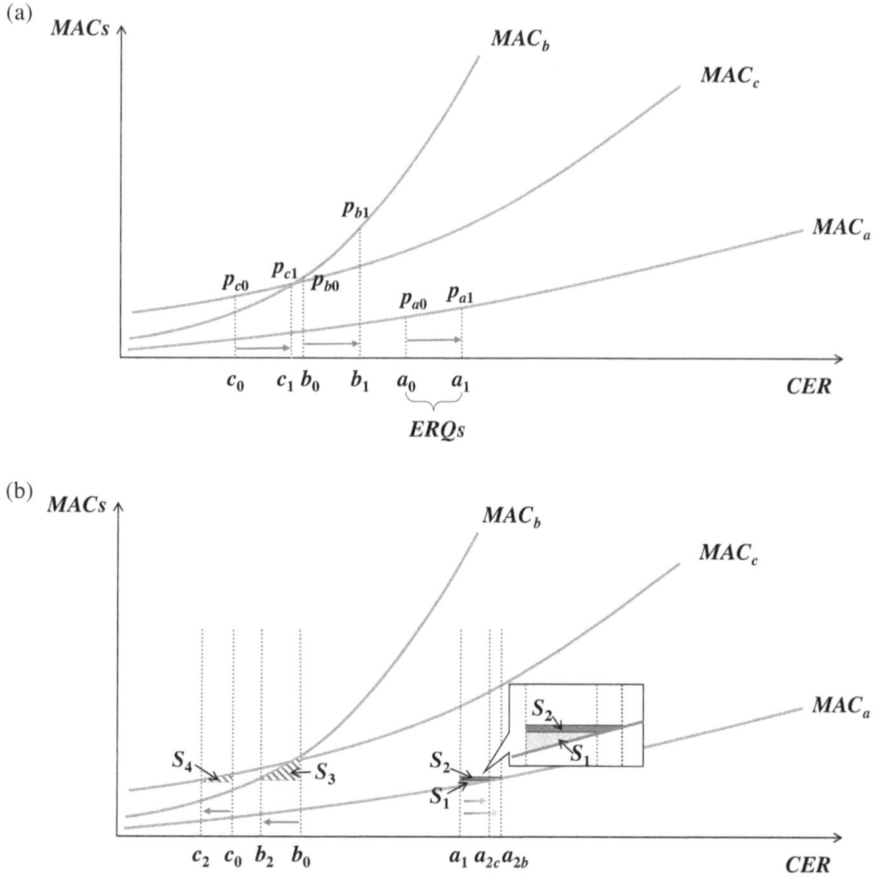

Figure 7.1 Schematic representation of varying marginal cost savings of three sectors

Note: **(a)** shows the case where only one sector is involved in the emission trading scheme (ETS). **(b)** shows the case where the second sector is selected into the ETS. *MACs* are marginal abatement costs, *CER* is carbon emission reduction, *ERQs* are emission reduction quotas, and MAC_a, MAC_b and MAC_c are the MAC curves for Sectors *a*, *b*, and *c*.

coverage of ETS. For example, suppose there are *n* ETS-covered sectors and *m* uncovered sectors, then we will have *m* new sets including *n+1* sectors, in which the potential MCSs of each newly covered sector are to be estimated.

Step 2: Allocate emission reduction quotas to ETS-covered sectors. For each new coverage set, we allocate the ERQs to each sector included in the ETS. Even though the selection of criteria for ERQs allocation has not yet reached an agreement, the cost-effectiveness of ETS can be independent from the allocation process (Chang et al., 2016; Zhou et al., 2013). Note that in the following we take the grandfathering scheme that allocates the ERQs to sectors based on the historical carbon emissions and that the comparison of various allocation schemes is out of our study.

Step 3: Estimate emission abatement costs in each new set. ETS that allows reallocation of ERQs between sectors can reduce EACs in achieving climate targets such as NDCs. To rank the ETS-covered sectors, we calculate the EACs in all the new sets and estimate the MCSs of each newly covered sector. By identifying the sector with the largest MCSs, the coverage of ETS from *n* sectors to *n*+1 sectors is determined.

Step 4: Iterate Steps 1–3 until all the sectors are ranked. Now that we have a newly determined sectoral coverage with *n*+1 sectors, we successively iterate Steps 1–3 until all the sectors are ranked in order when they are included into the ETS. In this sense, the COASCO method makes it possible to minimize the total EACs associated with any sectoral coverages, strictly conforming with the pursuit of cost-effectiveness in the ETS design.

In the following, we elaborate on the COASCO method with an example of China.

7.2.1 Measuring China's total emission reduction quotas by 2030

In keeping with China's NDCs of carbon intensity reduction by 2030, we follow Mi et al. (2017) and Fang et al. (2021) who assume an exponential decline in carbon intensity. The historical trend in China's carbon intensity that decreases from 2.96 tons per 10 thousand RMB in 2005 to 1.61 tons per 10 thousand RMB in 2017 fits well with our exponential model. Moreover, considering the uncertainty of China's economic growth in a post-COVID-19 'new normal' (Song & Zhou, 2020), we set three scenarios for future economic growth rate (i.e., 3%, 4%, and 5%) until 2030. Note that this is generally in line with the China's latest economic plan that requires the average annual GDP growth rate to be 4.7% consistently in the next 15 years in achieving the goal of becoming a moderately developed country by 2035 (Asian Development Bank 2021; National Development and Reform Commission of China 2021). The measurement of China's ERQs can be done as:

$$m = 1 - \sqrt[13]{\frac{CI_{2030}}{CI_{2017}}} = 1 - \sqrt[13]{\frac{CI_{2005} \times (1 - 65\%)}{CI_{2017}}} \tag{7.1}$$

$$CI_t = CI_{2017} \times (1 - m)^{t-2017} \tag{7.2}$$

$$ERQs_t = GDP_t \times (CI_{t-1} - CI_t) \tag{7.3}$$

$$GDP_t = GDP_{2017} \times (1 + \alpha)^{t-2017} \tag{7.4}$$

where *m* is the mitigation rate (%) of China's carbon intensity from 2018 to 2030. Given China's updated NDCs, we presume a 65% reduction of carbon intensity in 2030 against the 2005 levels. CI_t is China's carbon intensity (tons CO_2 per 10 thousand RMB) in year *t*, $ERQs_t$ is China's emission reduction quotas (million tons, Mt) in year t, GDP_t is China's gross domestic product (100 million RMB) in year t, and α is China's economic growth rate (%) from 2018 to 2030.

7.2.2 Allocating the emission reduction quotas to ETS-covered sectors

The total ERQs by 2030 will be allocated to ETS-covered sectors through a top-down approach. Admittedly, there are different schemes for ERQs allocation at the sectoral scale (e.g., grandfathering scheme, output-based scheme), and a consensus on the most appropriate one has not been reached (Fang et al., 2019; Lucas et al., 2020). However, the cost-effectiveness of ETS can be independent from the allocation process (Chang et al., 2016; Zhou et al., 2013). In this article, we adopt the grandfathering scheme that allocates ERQs proportionally to the sectoral historical emissions, since it makes the allocation more implementable than other schemes, particularly at the early stage of ETS (Cramton & Kerr, 2002; Wang et al., 2019). The ERQs allocation based on the grandfathering scheme can be done as:

$$CEA_t^i = \left(\sum_i CE_t^i - ERQs_t\right) \cdot \frac{CE_t^i}{\sum_i CE_t^i} \tag{7.5}$$

$$ERQs_t^i = CE_t^i - CEA_t^i \tag{7.6}$$

$$ERQs_t^i = ERQs_t \cdot \frac{CE_{t_0}^i}{\sum_i CE_{t_0}^i} \tag{7.7}$$

where CEA_t^i is the carbon emission allowance (Mt) of sector I in year t, CE_t^i is the carbon emissions (Mt) of sector i in year t, $ERQs_t^i$ is the emission reduction quotas (Mt) of sector i in year t, and t_0 is the baseline year. As seen in Formula 5, we allocate the CEA to each sector by multiplying China's total CEA and the proportion of sectoral emissions in each year t. Then to simulate the ETS in terms of MACs, we translate CEA to ERQs as in previous studies (Fang et al., 2021; Zhou et al., 2013). In theory, CEA represents the emissions that agents must comply with, and ERQs represent the emissions that should be reduced to meet the caps. In this sense, the translation will have no impacts on the EACs and carbon price. Specifically, the sectoral ERQs then can be calculated by subtracting the allocated CEA from its future carbon emissions (Formula 6). We assume that the proportion of carbon emissions by each sector keeps unchanged from 2018 to 2030 for brevity. While this probably increases the uncertainty of the estimated ERQs by sector, it has nothing related to our simulation on the impacts of sectoral coverage on the total EACs. In doing so, Formulas 5 and 6 can be converted to Formula 7 that accounts for the sectoral ERQs proportional to the historical emissions of each sector. Here the historical emissions refer to the annual average carbon emissions between 2015 and 2017.

7.2.3 Estimating the marginal abatement cost curves for sectors

The MAC curves depict how the EACs of a sector increase due to an additional unit of CER. Given the growing MACs when CER increases, we adopt a second-order polynomial function as a basis for estimating the MAC curves for sectors, as suggested by previous studies (Cui et al., 2014; Fan et al., 2016; Zhou et al., 2013). Specifically, we follow Zhou *et al.* (2013) and use accumulative carbon emission reduction (defined in this paper as the CER added up between 2000 and the specific year) as a proxy for the emission reduction potential. To account for the MACs, we introduce the potential economic loss proposed by Fang *et al.* (2021) as a substitute for the investment on clean development mechanism. Potential economic loss that calculates the economic loss due to the production inefficiency and accounts for the direct and indirect costs of CER is expected to correct the serious distortion in the MACs estimation given the phaseout of clean development mechanism since 2008 (Maraseni, 2013; World Bank 2012). The MAC curves for sectors are estimated as:

$$PEL_t^i = a^i \cdot (ACER_t^i)^2 + b^i \cdot ACER_t^i + c^i \tag{7.8}$$

$$PEL_t^i = \left(EP_{t+1}^i - EP_t^i\right) \cdot CE_t^i \tag{7.9}$$

$$ACER_t^i = \sum_{t_0}^{t} CER_t^i = \sum_{t_0}^{t} (CI_{t-1}^i - CI_t^i) \cdot GDP_t^i \tag{7.10}$$

where PEL_t^i is the potential economic loss (100 million RMB) of sector i in year t, $ACER_t^i$ is the accumulative carbon emission reduction (Mt) of sector i in year t while CER_t^i is the carbon emission reduction (Mt) of sector i in the single year t. In this study, we set 2000 as the starting year for measuring the cumulative value. EP_t^i is the emission productivity (10 thousand RMB per ton CO_2) of sector i in year t, expressed as the inverse of carbon intensity, and a_i, b_i and c_i are parameters to be estimated. Specifically, to account for the three parameters for each sector, we calculate potential economic loss and accumulative CER for all the sectors from 2000 to 2017 by Formulas 9 and 10. As shown in Formula 9, the potential economic loss of sectors is calculated by multiplying the difference in emission productivity and carbon emissions between adjacent years.

7.2.4 Modeling the emission trading scheme through nonlinear programming

In the ETS regime, the sectors with higher MACs tend to transfer their ERQs to those that have lower MACs. By reallocating the ERQs to sectors with lower MACs, the ETS enables the rigorous and cost-effective achievement of climate targets. We develop a non-linear programming model to simulate China's inter-sectoral ETS, and estimate China's EACs under different sectoral coverages, following the equations below:

$$\min TAC = \sum_i \int_{ACER^i_{2018}}^{ACER^i_{2030}} PEL^i_t \, dACER^i_t \tag{7.11}$$

$$ACER^i_{2030} - ACER^i_{2017} - Trade_i \geq \sum_{t=2018}^{2030} ERQs^i_t \tag{7.12}$$

$$\sum_i Trade_i = 0 \tag{7.13}$$

where *TAC* is China's total abatement costs (100 million RMB) after trading, and $Trade_i$ is the trading ERQs (Mt) of sector *i* from 2018 to 2030. Note that this study concerns the ERQs trade between sectors instead of CEA trade, whereby the positive $Trade_i$ refers to additional emissions reduced by Sector *i* for others, and vice versa. Equation 7.11 sets the objective function of minimizing China's EACs in 2018–2030. Equation 7.12 implies that the emissions reduced by each sector in 2018–2030 and the trading ERQs should not be fewer than its initially assigned climate targets. Equation 7.13 suggests a market clearing of ETS in this period.

7.2.5 Data

As shown in Table 7.1, the Chinese economy is classified into 29 sectors, of which 24 are sub-sectors in the manufacturing industry. Because there is no official carbon intensity published at national and sectoral scales, we calculate these values by dividing the carbon emissions by the corresponding GDP. The emission data for China and all the sectors in 2000–2017 come from the China Emission Accounts and Datasets. Since this is an emission inventory of 45 sectors from 1997 to 2017, we aggregate these data according to our sectoral classification. The GDP data for China and all the sectors in 2000–2017 are derived from the Chinese Statistical Yearbooks, all of which have been converted to the 2005 prices for comparability. Due to the lack of access to GDP of sub-sectors in the manufacturing industry, we use the data on value added of industrial enterprises above designed size to estimate the GDP of each sub-sector.

7.3 Results

7.3.1 Marginal abatement cost curves for sectors

Based on the sectoral potential economic loss and accumulative CER from 2000 to 2017, we estimate the MAC curves for 29 sectors using the quadratic function as displayed in Figure 7.2. In general, the MACs experience an exponential rise as

Table 7.1 Sectors and their abbreviations

Code	*Sector (Abbreviation)*	*Code*	*Sector (Abbreviation)*
1	Agriculture (*Agr*)	16	General machinery (*GenMac*)
2	Coal mining (*CoalMin*)	17	Specialist machinery (*SpecMac*)
3	Petroleum and gas (*Petr&Gas*)	18	Transport equipment (*TranEqp*)
4	Metal mining (*MetlMin*)	19	Electrical equipment (*ElecEqp*)
5	Nonmetal mining (*NonmetlMin*)	20	Electronic equipment (*ElenEqp*)
6	Food processing and tobaccos (*Food&Tob*)	21	Instrument and meter (*Inst&Metr*)
7	Textile (*Textile*)	22	Other manufacturing (*OthrMan*)
8	Clothing, leather, fur, etc. (*Clothing*)	23	Electricity and hot water production and supply (*E&W Prod*)
9	Wood processing and furnishing (*Wood&Furn*)	24	Gas production and supply (*GasProd*)
10	Paper making, printing, stationery, etc. (*Paper*)	25	Water production and supply (*WatProd*)
11	Petroleum refining, coking, etc. (*Petr&Coking*)	26	Construction (*Constr*)
12	Chemical industry (*ChemInd*)	27	Wholesale and retailing (*Whol&Ret*)
13	Nonmetal products (*NonmetlProd*)	28	Transport and storage (*Tran&Stor*)
14	Metallurgy (*Metl*)	29	Other services (*OthrServ*)
15	Metal products (*MetlProd*)		

the sectoral CER increases. Specifically, *GenMac*, *Whol&Ret* and *ElecEqp* have high MACs irrespective of CER, as they are less carbon-intensive and thus have little potential of CER. By contrast, for those large emitters (e.g., *E&W Prod*, *Metl*, *NonmetlProd*), the MACs always remain at a low level regardless of the CER. In addition, the MACs of *Tran&Stor*, *ChemInd*, *CoalMin* start at the bottom but experience a sharp increase in MACs in response to the deep emission reduction. For

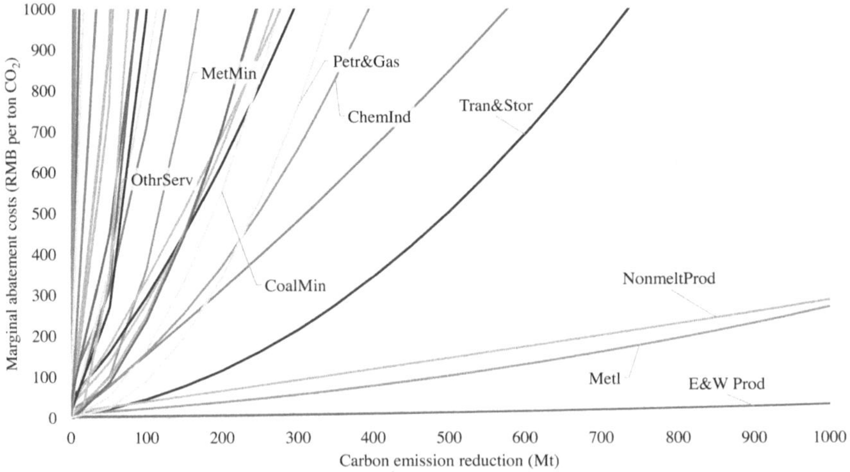

Figure 7.2 The marginal abatement cost curves for 29 sectors

instance, when the CER is assumed at 50 Mt, *Tran&Stor*, *ChemInd*, and *CoalMin* have their MACs at 22.44, 25.22, and 53.42 RMB per ton. But when the CER increases to 250 Mt, an additional unit of CER of these three sectors will cost 161.32, 504.94, and 829.16 RMB. In addition to the varying MAC curves for sectors, the difference in the historical CER between sectors is pronounced. For instance, *NonmetlProd*, *E&W Prod*, and *Metl* with relatively flat MAC curves have historical CER over 500 Mt, while the majority of the sectors (e.g., *Whol&Ret*, *Petr&Coking*, *CoalMin*) have their historical CER less than 100 Mt.

In theory, the varying sectoral MACs provide a basis for implementing the ETS across sectors. However, despite the considerable divergence of sectoral MACs, we notice some sectors that have similar MAC curves, e.g. *Metl* and *NonmetlProd*, *Tran&Stor* and *Petr&Coking*, *Petr&Gas* and *ChemInd*. Covering the sectors resembling in their MAC curves might contribute little to the potential cost savings of ETS while leading to a substantial increase in the administrative and transaction costs. With the aim of reducing more EACs with a relatively small sectoral coverage, there is a great need to identify the prioritized sectors that have large MCSs.

7.3.2 China's emission abatement costs and carbon price

Figure 7.3 shows that China's carbon intensity has reduced from 2.96 tons per 10 thousand RMB in 2005 to 1.61 tons per 10 thousand RMB in 2017, and will continuously decline to 1.04 tons per 10 thousand RMB in 2030 at a mitigation rate of

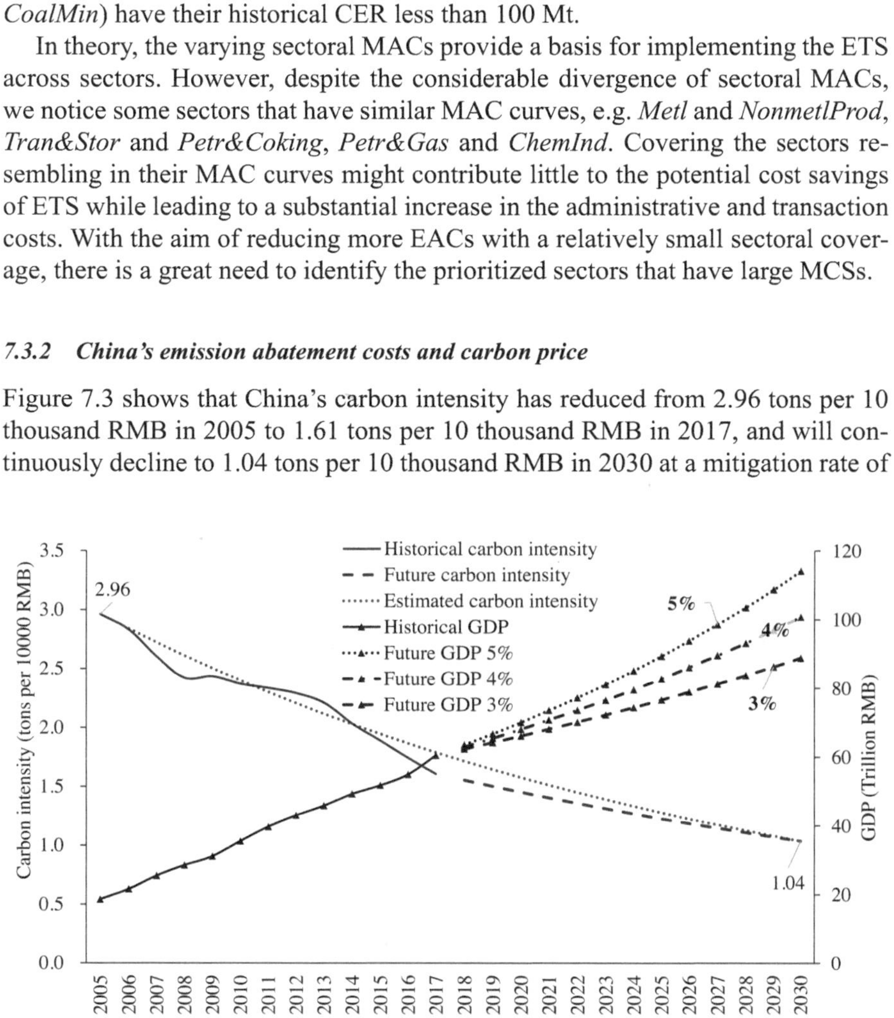

Figure 7.3 The historical and future trends of China's carbon intensity and GDP

Note: The black solid curve shows the historical carbon intensity from 2005 to 2017, the black dashed curve shows the future carbon intensity from 2018 to 2030, and the black dotted curve estimates the trajectory of China's carbon intensity from 2005 to 2030 in an exponential decline. Likewise, the solid-triangle curve shows the historical GDP from 2005 to 2017, and the triangle curves in dashed, dot-dashed and dotted project China's GDP at an economic growth rate of 3%, 4%, and 5%, respectively.

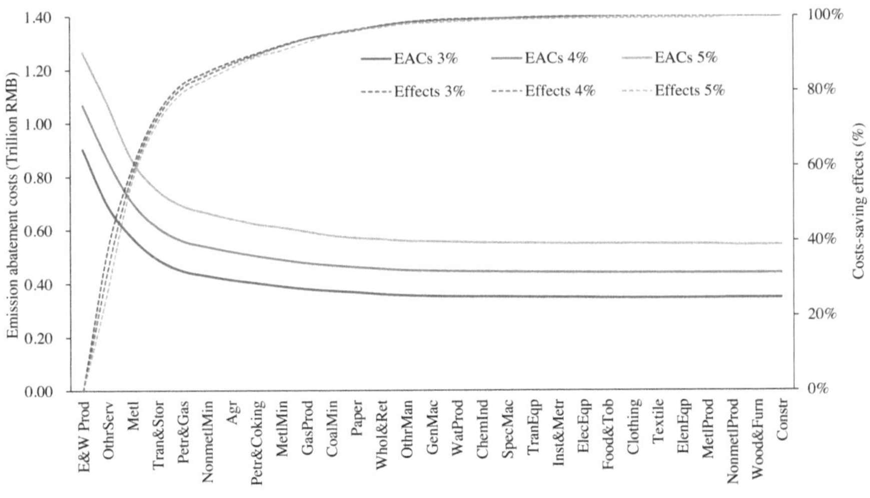

Figure 7.4 The changes in China's emission abatement costs (EACs) and marginal cost savings with different sectoral coverages

Note: Axis *X* from left to right shows the rankings of sectors to be covered in the emission trading scheme. The curves in dark, medium, and light gray show the trajectories at an economic growth rate of 3%, 4%, and 5%, respectively. The solid curves show China's EACs in different sectoral coverages, and the slashed curves show to what extent China's EACs can be saved as a result of coverage extension.

3.3% according to our exponential model. After projecting China's future GDP in the three scenarios for economic growth (i.e., 3%, 4%, and 5%), we measure China's total ERQs from 2018 to 2030 that amount to 4217.78, 4513.30, and 4832.60 Mt, respectively. As seen, enhancing the economic growth will moderately lead to more ERQs in the next decade.

By estimating and comparing the MCSs of sectors, the rankings of ETS-covered sectors can be determined. Figure 7.4 shows the impacts of each sector on China's EACs when it is involved in the ETS. In general, China's EACs can be largely reduced due to coverage extension. As seen, at an economic growth rate of 3%, 4%, and 5%, the total EACs will drop from 0.90 to 0.35, 1.07 to 0.44, and 1.27 to 0.55 trillion RMB, respectively. Meanwhile, we notice a sharp decrease in the MCSs as the sectoral coverage extends. Regardless of the economic growth scenarios, China's EACs can be cut by over 80% with the engagement of six sectors (i.e., *E&W Prod*, *OthrServ*, *Metl*, *Tran&Stor*, *Petr&Gas,* and *NonmetlMin*) compared to the engagement of *E&W Prod* only. Alternatively, the other 23 sectors contribute less than 20% to reducing the total EACs. This intriguing observation that a small sectoral coverage can save large EACs in the ETS conforms with the Pareto principle that 80% of effects arise from 20% of the causes (also referred to as the 80/20 rule).

Likewise, our simulation of China's MACs (i.e., the equilibrium carbon price) supports the Pareto principle as well (Figure 7.5). By including the first six sectors (i.e., *E&W Prod*, *OthrServ*, *Metl*, *Tran&Stor*, *Petr&Gas*, and *NonmetlMin*), China's MACs can be cut by around 80%, from 802.25, 895.13 and 1001.22 RMB

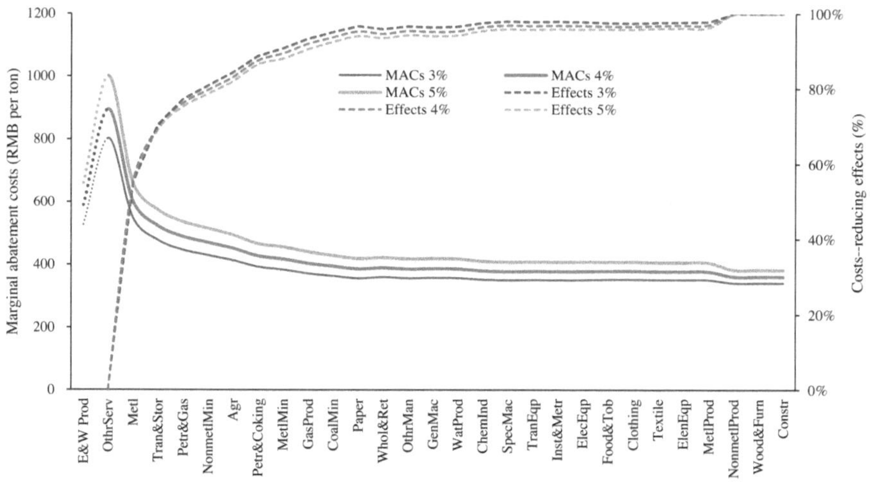

Figure 7.5 The changes in China's marginal abatement costs (MACs) and marginal cost reduction with different sectoral coverages

Note: The curves in dark, medium, and light gray show the trajectories at an economic growth rate of 3%, 4%, and 5%, respectively. The solid curves show China's MACs in different sectoral coverages, and the slashed curves show to what extent China's MACs can be reduced as a result of coverage extension.

per ton to 341.44, 261.41, and 383.49 RMB per ton at an economic growth rate of 3%, 4%, and 5%, respectively. Note that the inter-sectoral trading is not allowed until two or more sectors are engaged. While including *OthrSev* in the ETS might increase China's MACs compared to the case where only *E&W Prod* is engaged, it can leave the tertiary industry more room for economic growth and reduce the total EACs considerably. Moreover, we elaborate on the reallocation of the ERQs between sectors through ETS, whereas the technical countermeasures of each sector to realize climate targets (e.g., investment on low-carbon technology, transition of energy mix) are out of this study.

7.3.3 Sectoral emission reduction quotas and abatement costs

Sectoral coverage plays a critical role in the distribution of initial ERQs. As shown in Figure 7.6, under the grandfathering scheme, more initial ERQs tend to be assigned to the sectors that have large historical emissions, e.g., *E&W Prod*, *Metl*, *Tran&Stor*. Overall, the wider coverage of sectors can lead to a more even distribution of initial ERQs across sectors. For example, including the first four sectors (i.e., *E&W Prod*, *OthrServ*, *Metl*, and *Tran&Stor*) in the ETS considerably reduces the ERQs share assigned to *E&W Prod* and *Metl* from 100% and 29.6% to 60.3% and 26.5%, respectively, and this ERQs distribution remains largely steady when more sectors are engaged.

Figure 7.7 shows the impacts of sectoral coverage on the ERQs traded between sectors. In general, the total trading volumes of ERQs between sectors

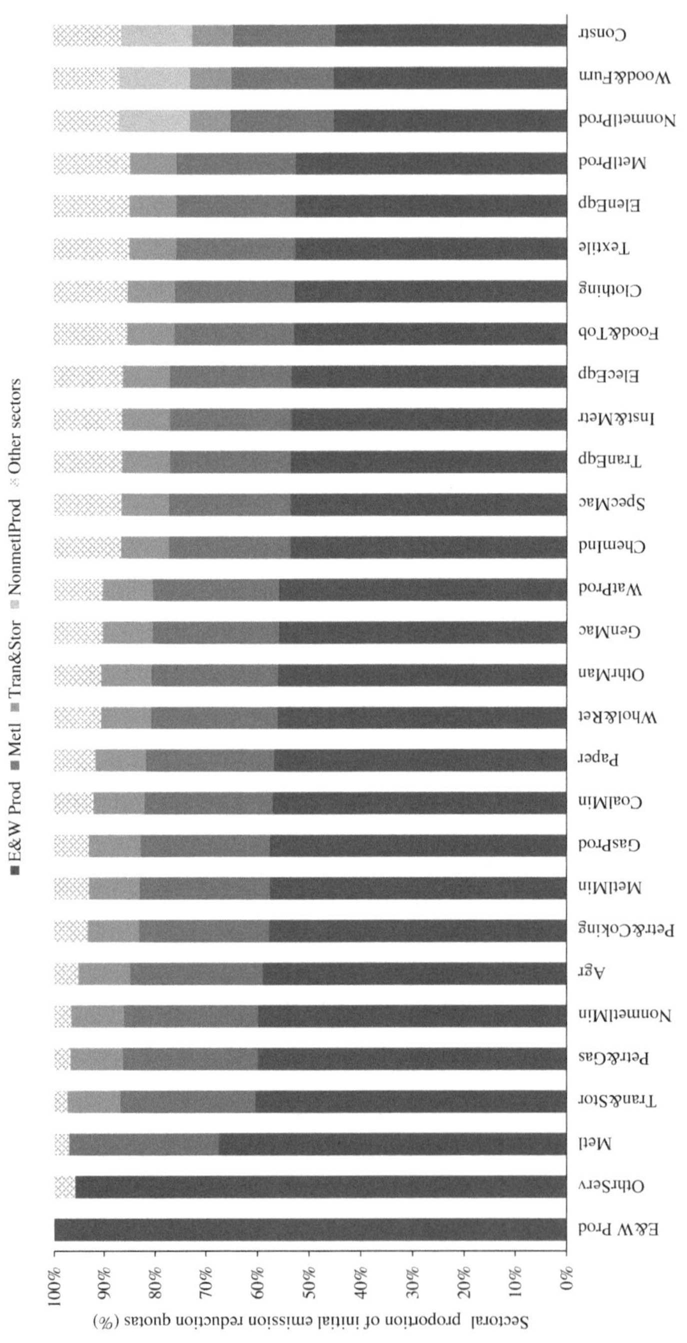

Figure 7.6 The distribution of initial emission reduction quotas (ERQs) assigned to sectors under different sectoral coverages

Note: We show the sectors having large initial ERQs and aggregate the rest with small ERQs into other sectors.

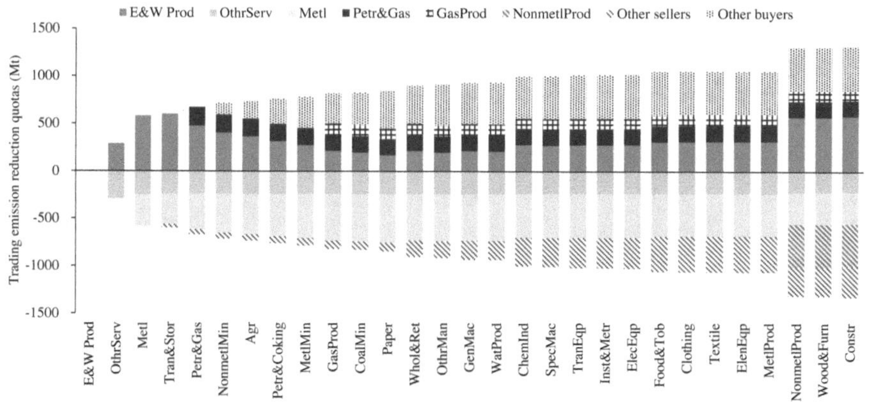

Figure 7.7 The trading emission reduction quotas (ERQs) between sectors under three economic growth scenarios (i.e., 3%, 4%, and 5%)

Note: The positive values refer to additional emissions reduced by one sector for others, and vice versa. We show those sectors having large ERQs trade and aggregate the rest into other sellers or other buyers.

experience a notable increase due to coverage extension that improves the market activation of ETS. Our simulation of three economic growth scenarios demonstrates that, on average, covering all the 29 sectors can increase the total trading volumes of ERQs from 289.78 Mt (when *E&W Prod* and *OthrServ* are involved) to 1320.41 Mt. However, the impacts of coverage extension on the trading ERQs vary considerably across sectors. For instance, in line with coverage extension, the trading ERQs of *E&W Prod* exhibit an N-shaped change while those of *Metl* exhibit a U-shaped change. Despite the benefits of market activation, the involvement of big ERQ traders runs the risk of market manipulation when the sectoral coverage extends (Bauer et al., 2020; Deng & Zhang, 2019; Wang et al., 2021). As such, we believe that covering the six sectors (i.e., *E&W Prod*, *OthrServ*, *Metl*, *Tran&Stor*, *Petr&Gas*, *NonmetlMin*) in the ETS can reach a balance between ETS market activation and risks, especially at its early stage.

We further calculate the ratio of sectoral EACs to GDP and explore the impacts of sectoral coverage on the EACs of sectors. Although the sectors involved in the ETS still have varying EACs, the differences between sectors can be reduced as the sectoral coverage extends. Specifically, our simulation result of three economic growth scenarios shows that, on average, *E&W Prod* and *Metl* will reduce the ratio of sectoral EACs to GDP from 1.3% to 0.1% and from 14.7% to 6.9% in response to the coverage extension, while the ratio for *Petr&Gas* increases from −2.7% to −1.7%. In other words, the EACs of sectors tend to converge as the sectoral coverage extends. However, covering the first ten sectors (e.g., *E&W Prod*, *OthrServ*, *Metl*) has enabled a convergence of EACs between sectors. Despite the same observation on cost convergence in all the three economic growth scenarios, we find that higher economic growth will increase the deviations of EACs between sectors,

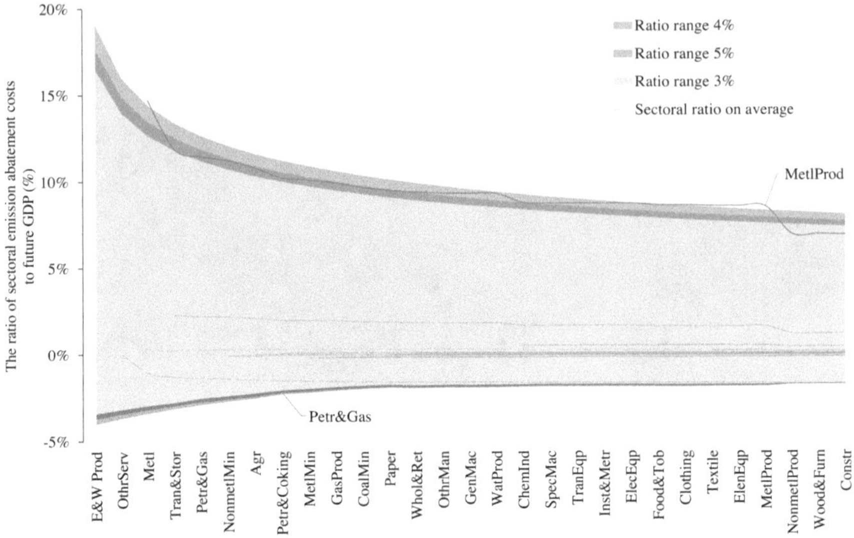

Figure 7.8 The ratio of sectoral emission abatement costs (EACs) to future GDP under different sectoral coverages

Note: The light gray shadow shows the ranges of ratio associated with different sectors at an economic growth rate of 3%, and the medium and dark gray shadows show the extended ranges at an economic growth rate of 4% and 5%, respectively. The gray curves show the average value (which refers to the ratio of EACs to future GDP) in the three economic growth scenarios for each of the 29 sectors as the sectoral coverage extends.

but this impact seems moderate because of the little difference between the three economic growth scenarios (Figure 7.8).

7.4 Discussion

7.4.1 Comparison with other criteria for sectoral coverage

The discussion over the criteria for sectoral coverage has emerged since the proposal of ETS. In addition to MCSs concerned in this paper, carbon intensity and trade intensity are two commonly adopted criteria focusing on emission reduction potential and carbon leakage, respectively (Qian et al., 2018; Wang et al., 2017). As said, the former refers to the carbon emissions per unit of GDP, and the latter refers to the ratio of the exports and imports to the total market scale. By estimating the MCSs of ETS-covered sectors under each criterion, we bring together these three criteria into Figure 7.9(a). Note that this comparison is performed at an economic growth rate of 4%, and the detailed sectoral rankings by carbon intensity or trade intensity can be found in Tables C.1 and C.2 in Appendix C. We show that *E&W Prod*, *Metl*, *NonmetlProd*, and *Tran&Stor* will be first included in the ETS if sectors are ranked by carbon intensity. However, compared to our study, the cost savings of ETS will remain at lower levels until *OthrServ* is engaged. While determining the sectoral coverage by trade intensity would allow

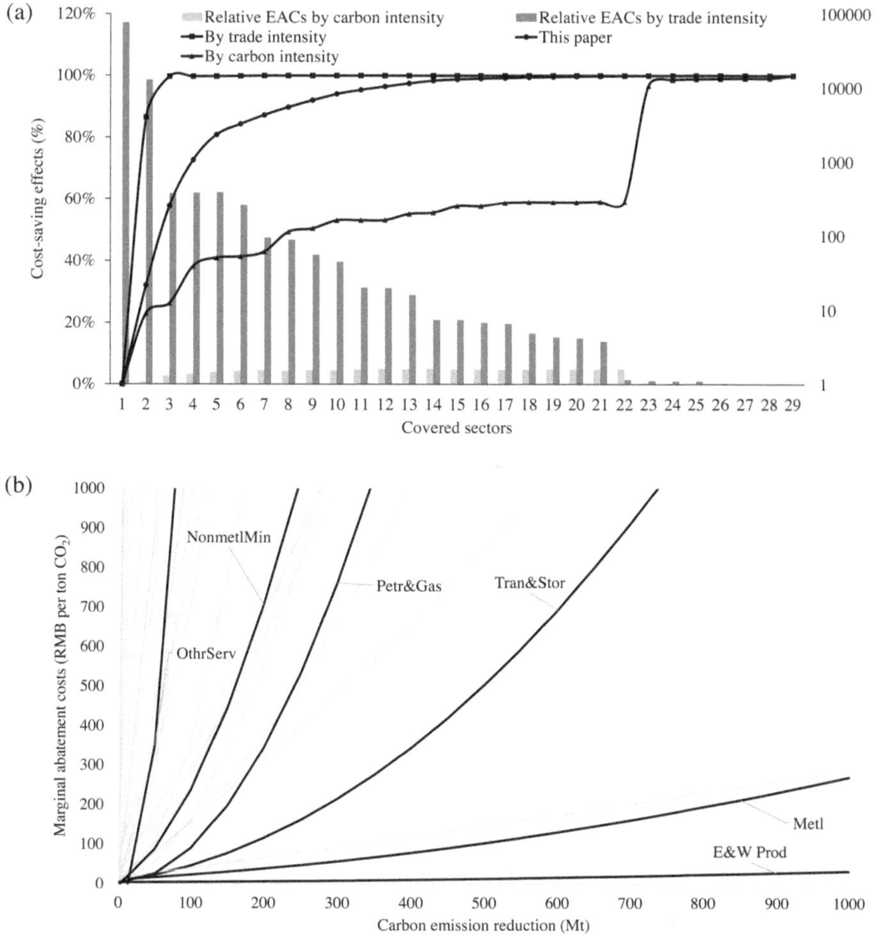

Figure 7.9 Cost savings of emission trading scheme (ETS) and emission abatement costs (EACs) by three criteria (a) and marginal abatement cost (MAC) curves for the first-covered sectors (b)

Note: (*a*) We rank 29 sectors by marginal cost savings (MCSs), carbon intensity, and trade intensity, and compare China's EACs under different criteria. The curves with squares, triangles, and circles show the cost savings of ETS when sectors are ranked by MCSs, carbon intensity, and trade intensity. The dark gray bars show the ratio of EACs by trade intensity to EACs by MCSs, and the light gray bars show the ratio of EACs by carbon intensity to EACs by MCSs. The ratios are expressed as a logarithm. (*b*) The black curves show the first six sectors (i.e., *E&W Prod*, *OthrServ*, *Metl*, *Tran&Stor*, *Petr&Gas*, *NonmetlMin*) to be covered in the ETS and the other 23 sectors are drawn in gray.

for soaring cost savings even with fewer ETS-covered sectors (e.g., *Clothing*, *ElenEqp*, *MetlProd*), it leads to excessively high EACs compared to the other two criteria because sectors having less emission reduction potential will be first included in the ETS. In summary, despite the insightful views on sectoral coverage from emission reduction potential and carbon leakage, our COASCO method is more appropriate in the sense that it allows for minimal EACs associated with any sectoral coverages.

By comparing the MCSs of sectors, our study recommends *E&W Prod*, *OthrServ*, *Metl*, *Tran&Stor*, *Petr&Gas* and *NonmetlMin* to be first included in the ETS. As shown in Figure 7.9(b), these six sectors differ considerably in their MAC curves that basically reflect the overall distribution of the MAC curves for all the 29 sectors. This implies that we should give priority to those sectors that have tremendously diverging MACs if the aim is to embrace the Pareto principle—a remarkable decline in EACs with a relatively small sectoral coverage. In other words, once the typical sectors that have large MCSs are identified, the Pareto principle can arise in terms of sectoral coverage.

7.4.2 Limitations and solutions

We acknowledge that there remain some limitations in the assumptions and methods of our analysis that need to be addressed in future studies. First, the use of macro data (e.g., sectoral carbon emissions, GDP) is likely to trigger an overestimation of the MACs by sector. Even though projecting future carbon price is not what this study mainly concerns, it is still desirable to further conduct a sensitivity analysis of sectoral MAC curves as long as the methodology underlying the MACs simulation becomes more scientifically robust. Second, while we set three scenarios to explore the impacts of economic growth on our empirical results, more socioeconomic factors that go beyond GDP should be considered at the next step. For example, China recently updates its NDCs in terms of carbon intensity reduction. More efforts to understand the impacts of this update on our empirical results would be required. Third, analysis of some highly aggregated sectors (e.g., *Agr*, *OthrServ*) in this paper may lead to impractical results. To strengthen the policy relevance of this study, a potential solution might be to present the results at the aggregated and disaggregated levels in keeping with the divergent requirements of policymakers and scientists. Fourth, we assume a long time period of carbon market clearing when all the agents are allowed for inter-temporal trading. Although this assumption may breach the reality of China's ETS and raise some uncertainties of the results, many researches have demonstrated that inter-temporal trading can smooth out the demand shocks of CEA and strengthen the resilience of ETS (Bocklet et al., 2019; Duan et al., 2018; Fankhauser & Hepburn, 2010). In practice, recent reform in the ETS of the European Union allows the inter-temporal trading by making use of the market stability reserve (Bocklet et al., 2019; Perino & Willner, 2017). Nevertheless, further analyses that divide long time period into several intervals can be the next step to examine the differentiated impacts of inter-temporal trading on the various agents. Fifth, in this article, we consider direct emissions of sectors merely and leave aside indirect emissions associated with outsourcing electricity requirements. While the negligence of indirect emissions might substantially increase the costs of electricity production that would be passed on to power consumers ultimately, some researches have demonstrated that the double counting that considers both direct and indirect emissions may cause a higher overall cap of ETS, thus leading to a lower emission reduction effect (Qian et al., 2018; Xiong et al., 2017; Zeng et al., 2018). Further analysis related to this issue would therefore be required. Last but not least, this paper focuses exclusively on ETS as a single policy tool for CER. By allocating all the ERQs to the ETS-covered sectors, the remaining sectors are excluded from CER. This may lead to unfairness and inefficiency because unengaged sectors would not reduce their carbon emissions proactively (Mu et al., 2018). Mixed policy packages including ETS, carbon tax, renewable energy subsidy, emission standards, and sophisticated emission accounting methods like carbon footprinting, remote sensing retrieval, machine learning, and big data analysis are expected to respond to this challenge (Duren & Miller, 2012; Gouveia & Palma, 2019; Liu & Lo, 2021; Long et al., 2019; Ye et al., 2021; Zhao et al., 2021).

7.5 Conclusions

For a policy that is primarily designed for minimizing the total EACs, previous studies dedicated to determining the sectoral coverage rarely concern the varying MCSs of sectors, thus undermining the cost-effective design of ETS. To estimate and compare the sectoral MCSs, this paper proposes the COASCO method that selects and ranks the sectors to be included in the ETS. Taking China's NDCs by 2030 as an example, we conduct an empirical study using this COASCO method to determine the rankings of 29 sectors to be covered in the ETS and explore the impacts of sectoral coverage on China's EACs under three economic growth scenarios. By providing a generalized and systematic framework for determining the sectoral coverage, this study makes it possible to minimize the total EACs associated with any sectoral coverages, thus being able to assist policymakers in designing a more cost-effective ETS that is the key to fulfilling China's latest ambitious goals of reaching carbon peaking by 2030 and carbon neutrality by 2060.

There are several key findings and policy implications from our simulation results. First, although the extension of sectoral coverage generally reduces China's EACs and improves the cost-effectiveness of ETS, the MCSs vary considerably between sectors and a small sectoral coverage can already lead to a substantial decline in the national EACs. Specifically, irrespective of the economic growth scenarios, covering the six sectors (i.e., *E&W Prod*, *OthrServ*, *Metl*, *Tran&Stor*, *Petr&Gas*, *NonmetlMin*) out of 29 can reduce China's EACs and MACs by over 80% compared to covering *E&W Prod* only, which conforms with the Pareto principle with respect to the sectoral coverage of ETS. Second, while coverage extension can somehow reduce the sectoral differences in EACs, complementary policies (e.g., fiscal transfers and subsidy, net ETS revenues recycling) that are necessary to further shrink the deviations of EACs between sectors would be highly needed. Third, despite the benefits of market activation as a result of coverage extension, the growing trading ERQs might give rise to big ERQs sellers and buyers (e.g., *E&W Prod*, *Metl*), probably increasing the risks of market manipulation. As per our simulation result, covering the first six sectors (i.e., *E&W Prod*, *OthrServ*, *Metl*, *Tran&Stor*, *Petr&Gas*, *NonmetlMin*) in the ETS is preferred as it allows for a balance between ETS market activation and risks. Last but not least, compared to the criteria of carbon intensity and trade intensity, ranking the ETS-covered sectors pertaining to the criterion of MCSs is more appropriate in ways it allows for minimization of the total EACs associated with any sectoral coverages, strictly conforming with the pursuit of cost-effectiveness in the ETS design.

References

Asian Development Bank. (2021). The 14th Five-Year Plan of the People's Republic of China—Fostering High-Quality Development. https://www.adb.org/sites/default/files/publication/705886/14th-five-year-plan-high-quality-development-prc.pdf.

Bauer, N., Bertram, C., Schultes, A., Klein, D., Luderer, G., Kriegler, E., Popp, A., & Edenhofer, O. (2020). Quantification of an efficiency–sovereignty trade-off in climate policy. *Nature*, *588*(7837), 261–266.

Baylis, K., Fullerton, D., & Karney, D. H. (2013). Leakage, welfare, and cost-effectiveness of carbon policy. *American Economic Review*, *103*(3), 332–337.

Bertram, C., Luderer, G., Pietzcker, R. C., Schmid, E., Kriegler, E., & Edenhofer, O. (2015). Complementing carbon prices with technology policies to keep climate targets within reach. *Nature Climate Change*, *5*(3), 235–239.

Betz, R., Sanderson, T., & Ancev, T. (2010). In or out: Efficient inclusion of installations in an emissions trading scheme? *Journal of Regulatory Economics*, *37*(2), 162–179.

Bistline, J. E., & Rose, S. K. (2018). Social cost of carbon pricing of power sector CO_2: Accounting for leakage and other social implications from subnational policies. *Environmental Research Letters*, *13*(1), 014027.

Bocklet, J., Hintermayer, M., Schmidt, L., & Wildgrube, T. (2019). The reformed EU ETS-intertemporal emission trading with restricted banking. *Energy Economics*, *84*, 104486.

Böhringer, C., Dijkstra, B., & Rosendahl, K. E. (2014a). Sectoral and regional expansion of emissions trading. *Resource and Energy Economics*, *37*, 201–225.

Böhringer, C., Fischer, C., & Rosendahl, K. E. (2014b). Cost-effective unilateral climate policy design: Size matters. *Journal of Environmental Economics and Management*, *67*(3), 318–339.

Chang, K., Zhang, C., & Chang, H. (2016). Emissions reduction allocation and economic welfare estimation through interregional emissions trading in China: Evidence from efficiency and equity. *Energy*, *113*, 1125–1135.

Convery, F. J. (2009). Origins and development of the EU ETS. *Environmental and Resource Economics*, *43*(3), 391–412.

Cramton, P., & Kerr, S. (2002). Tradeable carbon permit auctions: How and why to auction not grandfather. *Energy Policy*, *30*(4), 333–345.

Cui, L. B., Fan, Y., Zhu, L., & Bi, Q. H. (2014). How will the emissions trading scheme save cost for achieving China's 2020 carbon intensity reduction target? *Applied Energy*, *136*, 1043–1052.

Cui, R. Y., Hultman, N., Cui, D., McJeon, H., Yu, S., Edwards, M. R., Sen, A., Song, K., Bowman, C., Clarke, L., Kang, J., Lou, J., Yang, F., Yuan, J., Zhang, W., & Zhu, M. (2021). A plant-by-plant strategy for high-ambition coal power phaseout in China. *Nature Communications*, *12*(1), 1–10.

De Cara, S., & Jayet, P. A. (2011). Marginal abatement costs of greenhouse gas emissions from European agriculture, cost effectiveness, and the EU non-ETS burden sharing agreement. *Ecological Economics*, *70*(9), 1680–1690.

Deng, M., & Zhang, W. (2019). Recognition and analysis of potential risks in China's carbon emission trading markets. *Advances in Climate Change Research*, *10*(1), 30–46.

Di Maria, C., & Van der Werf, E. (2008). Carbon leakage revisited: Unilateral climate policy with directed technical change. *Environmental and Resource Economics*, *39*(2), 55–74.

Dong, F., Dai, Y., Zhang, S., Zhang, X., & Long, R. (2019). Can a carbon emission trading scheme generate the porter effect? Evidence from pilot areas in China. *Science of the Total Environment*, *653*, 565–577.

Duan, M., Qi, S., & Wu, L. (2018). Designing China's national carbon emissions trading system in a transitional period. *Climate Policy*, *18*(sup 1), 1–6.

Duren, R. M., & Miller, C. E. (2012). Measuring the carbon emissions of megacities. *Nature Climate Change*, *2*(8), 560–562.

Fang, K., Zhang, Q., Long, Y., Yoshida, Y., Sun, L., Zhang, H., Dou, Y., & Li, S. (2019). How can China achieve its intended nationally determined contributions by 2030? A multi-criteria allocation of China's carbon emission allowance. *Applied Energy*, *241*, 380–389.

Fang, K., Zhang, Q., Song, J., Yu, C., Zhang, H., & Liu, H. (2021). How can national ETS affect carbon emissions and abatement costs? Evidence from the dual goals proposed by China's NDCs. *Resources, Conservation and Recycling*, *171*, 105638.

Fankhauser, S., & Hepburn, C. (2010). Designing carbon markets. Part I: Carbon markets in time. *Energy Policy*, *38*(8), 4363–4370.

Fan, Y., & Wang, X. (2014). Which sectors should be included in the ETS in the context of a unified carbon market in China? *Energy and Environment*, *25*(3–4), 613–634.

Fan, Y., Wu, J., Xia, Y., & Liu, J. Y. (2016). How will a nationwide carbon market affect regional economies and efficiency of CO_2 emission reduction in China? *China Economic Review*, *38*, 151–166.

Fujimori, S., Kubota, I., Dai, H., Takahashi, K., Hasegawa, T., Liu, J. Y., Hijioka, Y., Masui, T., & Takimi, M. (2016). Will international emissions trading help achieve the objectives of the Paris Agreement? *Environmental Research Letters*, *11*(10), 104001.

Glanemann, N., Willner, S. N., & Levermann, A. (2020). Paris Climate Agreement passes the cost-benefit test. *Nature Communications*, *11*(1), 1–11.

Gouveia, J. P., & Palma, P. (2019). Harvesting big data from residential building energy performance certificates: Retrofitting and climate change mitigation insights at a regional scale. *Environmental Research Letters*, *14*(9), 095007.

Green, J. F. (2021). Does carbon pricing reduce emissions? A review of ex-post analyses. *Environmental Research Letters*, *16*(4), 043004.

Heggelund, G., Stensdal, I., Duan, M., & Wettestad, J. (2019). China's development of ETS as a GHG mitigating policy tool: A case of policy diffusion or domestic drivers? *Review of Policy Research*, *36*(2), 168–194.

IPCC (2014). *Climate change 2014: Synthesis report*. IPCC.

Jarke, J., & Perino, G. (2017). Do renewable energy policies reduce carbon emissions? On caps and inter-industry leakage. *Journal of Environmental Economics and Management*, *84*, 102–124.

Jiang, J., Xie, D., Ye, B., Shen, B., & Chen, Z. (2016). Research on China's cap-and-trade carbon emission trading scheme: Overview and outlook. *Applied Energy*, *178*, 902–917.

Juergens, I., Barreiro-Hurlé, J., & Vasa, A. (2013). Identifying carbon leakage sectors in the EU ETS and implications of results. *Climate Policy*, *13*(1), 89–109.

Karstensen, J., & Peters, G. (2018). Distributions of carbon pricing on extraction, combustion and consumption of fossil fuels in the global supply-chain. *Environmental Research Letters*, *13*(1), 014005.

Lamb, W. F., Wiedmann, T., Pongratz, J., Andrew, R., Crippa, M., Olivier, J. G., Wiedenhofer, D., Mattioli, G., Al Khourdajie, A., House, J., Pachauri, S., Figueroa, M., Saheb, Y., Slade, R., Hubacek, K., Sun, L., Ribeiro, S. K., Khennas, S., de la Rue du Can, S., Chapungu, L., Davis, S. J., Bashmakov, I., Dai, H., Dhakal, S., Tan, X., Geng, Y., Gu, B., & Minx, J. C. (2021). A review of trends and drivers of greenhouse gas emissions by sector from 1990 to 2018. *Environmental Research Letters*, *16*(7), 073005.

Lin, B., & Jia, Z. (2019). What will China's carbon emission trading market affect with only electricity sector involvement? A CGE based study. *Energy Economics*, *78*, 301–311.

Lin, B., & Jia, Z. (2020). Why do we suggest small sectoral coverage in China's carbon trading market? *Journal of Cleaner Production*, *257*, 120557.

Liu, L., Chen, C., Zhao, Y., & Zhao, E. (2015a). China's carbon-emissions trading: Overview, challenges and future. *Renewable and Sustainable Energy Reviews*, *49*, 254–266.

Liu, Z., Guan, D., Moore, S., Lee, H., Su, J., & Zhang, Q. (2015b). Climate policy: Steps to China's carbon peak. *Nature*, *522*(7556), 279.

Liu, M., & Lo, K. (2021). Pathways to international cooperation on climate governance in China: A comparative analysis. *Journal of Chinese Governance*, *6*(3), 417–434.

Long, Y., Yoshida, Y., Fang, K., Zhang, H., & Dhondt, M. (2019). City-Level household carbon footprint from purchaser point of view by a modified input-output model. *Applied Energy*, *236*, 379–387.

Lucas, P. L., Wilting, H. C., Hof, A. F., & van Vuuren, D. P. (2020). Allocating planetary boundaries to large economies: Distributional consequences of alternative perspectives on distributive fairness. *Global Environmental Change*, *60*, 102017.

Lu, Q., Fang, K., Heijungs, R., Feng, K., Li, J., Wen, Q., Li, Y., & Huang, X. (2020). Imbalance and drivers of carbon emissions embodied in trade along the Belt and Road Initiative. *Applied Energy*, *280*, 115934.

Maraseni, T. N. (2013). Selecting a CDM investor in China: A critical analysis. *Energy Policy*, *53*, 484–489.

Mi, Z., Wei, Y. M., Wang, B., Meng, J., Liu, Z., Shan, Y., Liu, J., & Guan, D. (2017). Socioeconomic impact assessment of China's CO_2 emissions peak prior to 2030. *Journal of Cleaner Production*, *142*, 2227–2236.

Mu, Y., Evans, S., Wang, C., & Cai, W. (2018). How will sectoral coverage affect the efficiency of an emissions trading system? A CGE-based case study of China. *Applied Energy*, *227*, 403–414.

National Development and Reform Commission of China. (2021). Outline of the People's Republic of China 14th Five-Year Plan for National Economic and Social Development and Long-Range Objectives for 2035. https://www.ndrc.gov.cn/xxgk/zcfb/ghwb/202103/P020210323538797779059.pdf.

Perdan, S., & Azapagic, A. (2011). Carbon trading: Current schemes and future developments. *Energy Policy*, *39*(10), 6040–6054.

Perino, G., & Willner, M. (2017). EU-ETS phase IV: Allowance prices, design choices and the market stability reserve. *Climate Policy*, *17*(7), 936–946.

Qian, H., Zhou, Y., & Wu, L. (2018). Evaluating various choices of sector coverage in China's national emissions trading system (ETS). *Climate Policy*, *18*(sup1), 7–26.

Qi, S., & Cheng, S. (2018). China's national emissions trading scheme: Integrating cap, coverage and allocation. *Climate Policy*, *18*(sup1), 45–59.

Skjærseth, J. B., & Wettestad, J. (2008). Implementing EU emissions trading: Success or failure? *International Environmental Agreements: Politics, Law and Economics*, *8*(3), 275–290.

Song, L., & Zhou, Y. (2020). The COVID-19 pandemic and its impact on the global economy: What does it take to turn crisis into opportunity. *China & World Economy*, *28*(4), 1–25.

Spash, C. L. (2010). The brave new world of carbon trading. *New Political Economy, 15*(2), 169–195.

Springmann, M. (2012). A look inwards: Carbon tariffs versus internal improvements in emissions-trading systems. *Energy Economics, 34*, S228–S239.

Stoerk, T., Dudek, D. J., & Yang, J. (2019). China's national carbon emissions trading scheme: Lessons from the pilot emission trading schemes, academic literature, and known policy details. *Climate Policy, 19*(4), 472–486.

Tan, X., Liu, Y., Cui, J., & Su, B. (2018). Assessment of carbon leakage by channels: An approach combining CGE model and decomposition analysis. *Energy Economics, 74*, 535–545.

Tang, B. J., Ji, C. J., Hu, Y. J., Tan, J. X., & Wang, X. Y. (2020). Optimal carbon allowance price in China's carbon emission trading system: Perspective from the multi-sectoral marginal abatement cost. *Journal of Cleaner Production, 253*, 119945.

Wang, X., Teng, F., Wang, G., Zhou, S., & Cai, B. (2018a). Carbon leakage scrutiny in ETS and non-ETS industrial sectors in China. *Resources, Conservation and Recycling, 129*, 424–431.

Wang, X., Teng, F., Zhou, S., & Cai, B. (2017). Identifying the industrial sectors at risk of carbon leakage in China. *Climate Policy, 17*(4), 443–457.

Wang, X., Zhang, X. B., & Zhu, L. (2019). Imperfect market, emissions trading scheme, and technology adoption: A case study of an energy-intensive sector. *Energy Economics, 81*, 142–158.

Wang, X., Zhu, L., & Fan, Y. (2018b). Transaction costs, market structure and efficient coverage of emissions trading scheme: A microlevel study from the pilots in China. *Applied Energy, 220*, 657–671.

Wang, X., Zhu, L., & Liu, P. (2021). Manipulation via endowments: Quantifying the influence of market power on the emission trading scheme. *Energy Economics, 103*, 105533.

Weikmans, R., Asselt, H. V., & Roberts, J. T. (2020). Transparency requirements under the Paris Agreement and their (un) likely impact on strengthening the ambition of nationally determined contributions (NDCs). *Climate Policy, 20*(4), 511–526.

Weitzman, M. L., & Spence, A. M. (1978). *Regulatory strategies for pollution control.* M.I.T. Press.

World Bank. (2012). State and trends of the carbon market 2012.

Wu, L., Zhang, S., & Qian, H. (2022a). Distributional effects of China's national emission trading scheme with an emphasis on sectoral coverage and revenue recycling. *Energy Economics, 105*, 105770.

Wu, L., Zhou, Y., & Qian, H. (2022b). Global actions under the Paris agreement: Tracing the carbon leakage flow and pursuing countermeasures. *Energy Economics, 106*, 105804.

Xie, Y., Wu, Y., Xie, M., Li, B., Zhang, H., Ma, T., & Zhang, Y. (2020). Health and economic benefit of China's greenhouse gas mitigation by 2050. *Environmental Research Letters, 15*(10), 104042.

Xiong, L., Shen, B., Qi, S., Price, L., & Ye, B. (2017). The allowance mechanism of China's carbon trading pilots: A comparative analysis with schemes in EU and California. *Applied Energy, 185*, 1849–1859.

Ye, R., Gao, Z., Fang, K., Liu, K., & Chen, J. (2021). Moving from subsidy stimulation to endogenous development: A system dynamics analysis of China's NEVs in the post-subsidy era. *Technological Forecasting & Social Change, 168*, 120757.

Ying, J. J., & Sovacool, B. K. (2021). A fair trade? Expert perceptions of equity, innovation, and public awareness in China's future emissions trading scheme. *Climatic Change, 164*(3), 1–23.

Zeng, Y., Weishaar, S. E., & Vedder, H. H. (2018). Electricity regulation in the Chinese national emissions trading scheme (ETS): Lessons for carbon leakage and linkage with the EU ETS. *Climate Policy*, *18*(10), 1246–1259.

Zhang, Y. J., & Wei, Y. M. (2010). An overview of current research on EU ETS: Evidence from its operating mechanism and economic effect. *Applied Energy*, *87*(6), 1804–1814.

Zhang, Q., Wiedmann, T., Fang, K., Song, J., He, J., & Chen, X. (2022). Bridging planetary boundaries and spatial heterogeneity in a hybrid approach: A focus on Chinese provinces and industries. *Science of the Total Environment*, *804*, 150179.

Zhang, P., Yin, G., & Duan, M. (2020). Distortion effects of emissions trading system on intra-sector competition and carbon leakage: A case study of China. *Energy Policy*, *137*, 111126.

Zhao, Z., Yang, X., Yan, H., Huang, Y., Zhang, G., Lin, T., & Ye, H. (2021). Downscaling building energy consumption carbon emissions by machine learning. *Remote Sensing*, *13*(21), 4346.

Zhou, B., Zhang, C., Wang, Q., & Zhou, D. (2020). Does emission trading lead to carbon leakage in China? Direction and channel identifications. *Renewable and Sustainable Energy Reviews*, *132*, 110090.

Zhou, P., Zhang, L., Zhou, D. Q., & Xia, W. J. (2013). Modeling economic performance of interprovincial CO_2 emission reduction quota trading in China. *Applied Energy*, *112*, 1518–1528.

Zhou, B., Zhou, F., Zhou, D., Qiao, J., & Xue, B. (2021). Improvement of environmental performance and optimization of industrial structure of the Yangtze River economic belt in China: Going forward together or restraining each other? *Journal of Chinese Governance*, *6*(3), 435–455.

8 Co-benefits of Carbon Trading Scheme on Air Quality

8.1 Out-expected effects of emission trading scheme: Environmental and health co-benefits

Over the past decades, some parts of China have experienced severe air pollution, as exemplified by the particulate matter with a diameter <2.5 μm ($PM_{2.5}$) (Wang et al., 2017). The $PM_{2.5}$ concentrations in many Chinese cities exceed the National Ambient Air Quality Standards and the World Health Organization guideline (Fang et al., 2019). In responding to air pollution, the Chinese government has taken aggressive control measures since early this century, such as the sulfur dioxide (SO_2) control during the 11th Five Year Plan (2006–2010), and both SO_2 and nitrogen oxides (NO_x) control during the 12th Five Year Plan (2011–2015). Recently, more integrated policy actions, such as *the Air Pollution Prevention and Control Action Plan*, have been implemented in an efficient way (The State Council of China, 2013), resulting in a gradual decline in multiple atmospheric pollutants (Li et al., 2020).

When it comes to the global scale, an enormous amount of greenhouse gases (GHGs) and air pollutants associated with industrial activities has accelerated global warming and caused 5.6–6.6 million annual premature deaths (Fang et al., 2019; IPCC, 2017; Vandyck et al., 2018). Humanity is increasingly facing the challenges of reducing GHG emissions while combating air pollution (Lelieveld et al., 2015; Schmale et al., 2014). Climate change affects air quality through several pathways such as altering the ventilation and dilution of air pollutants, photochemical reactions, removal processes, biogenic emissions, and wildfires (Silva et al., 2017). Although the greenhouse effect of carbon dioxide (CO_2) emissions is global, the impact of air pollution can be highly heterogeneous and unequal, showing tremendous variation across regions and over time. Researchers have called for a coordinated effort to combat climate change and to improve air quality (Haines et al., 2017; Li et al., 2018).

Previous studies have explored the co-benefits of climate policies and air pollution control policies at multiple scales, demonstrating that CO_2 emissions are highly related to atmospheric pollutant emissions (Aunan et al., 2006; Dong et al., 2015; Rafaj et al., 2013), but also that controlling CO_2 emissions generally have positive effects on air quality and public health through reduced emissions of

DOI: 10.4324/9781003433859-8

co-pollutants, such as SO_2 and NO_x, (Haines et al., 2009). Earlier regional-specific estimates present a broad range of values for the co-benefits of climate policy on air quality (between 2 and 196 dollars per ton of CO_2) (Nemet et al., 2010). Recent work highlights that the improved human health outcomes due to cleaner air can largely offset the costs to reduce GHGs, particularly in heavily polluted regions (Thompson et al., 2014; West et al., 2013).

Emissions trading scheme (ETS) has been considered a key instrument for the mitigation of climate change by reducing CO_2 emissions (Liu et al., 2015). For those of the regions with serious air pollution, climate mitigation instruments (e.g., ETS) that have the potential to improve air quality would be favored by the public (Anderson et al., 2018; Cushing et al., 2018; Fang et al., 2020). Such co-benefits represent substantial cost savings that can be obtained by reducing CO_2 emissions. In this sense, the ETS can contribute to reshaping the spatial patterns of CO_2 emissions and associated pollutant emissions (Bain et al., 2016; Xie et al., 2018; Yu et al., 2015). Moreover, a phenomenon known as carbon leakage occurs if such difference in CO_2 emission costs triggers the relocation of carbon-intensive activities and related emissions from one region to another with less stringent climate policies (Wood et al., 2020; Zhang & Fang, 2019). Obviously, carbon leakage is not restricted to the international level and has been observed at multiple sub-national scales (Zeng et al., 2018). However, most studies on the co-benefits of ETS have made use of provincial or municipal data at the macro- and meso-level without consideration of corporate emissions (Sun et al., 2014). In other words, the environmental and health impacts of ETS have been rarely explored at the corporate level.

This limitation also applies to China—the world's largest energy consumer and GHGs emitter, where the ETS was originally implemented by seven pilots in 2011 and has gradually extended to a unified national trading system since 2017. Existing studies have concluded that the co-benefits of ETS are likely to be remarkable in some pilot areas in China (Aunan et al., 2006), such as Shanghai (Dong et al., 2015), Guangdong (Cheng et al., 2015), and Jing-Jin-Ji region (Cai et al., 2016). However, China's nationwide ETS is currently at an early stage, and its impact on the environment and human health remains underexplored. For instance, while health impacts associated with corporations that share the same emissions inventory may substantially disagree because of the discrepancy in local socioeconomic and natural conditions where the corporations are located (Fang et al., 2019; GEA, 2012), there is, to our understanding, a lack of analysis to show whether ETS will close the gap in carbon emissions among different regions and reduce atmospheric pollutants by tabulating, compiling and analyzing corporate-level data at a high spatial resolution (Zhang et al., 2019).

Hubei is the first ETS pilot in central China, whose industrial structure has much in common with the national average. As of 2015, Hubei's ETS comprised 163 corporations that can be divided into electricity, cement, steel, chemical engineering, nonferrous metal, automobile manufacturing, glass, paper, and other industries (HDRC, 2017). The trading volume of carbon emission allowances in Hubei amounted to 11.86 million tons and 270 million dollars in 2015, ranking first in the country, followed by Shenzhen and Guangdong, whose trading volume accounted only for 67% and 71% of that of Hubei, respectively (CHEE, 2015).

To fill in the gap in our knowledge, this study investigated the trading of corporate emission allowances in Hubei by tracking the changes in the spatial distribution of CO_2 and $PM_{2.5}$ with and without ETS and estimating the cause-specific mortality associated with air pollution. To that end, we incorporated corporate-level CO_2 and co-pollutant emissions data with a high spatial resolution (5 km × 5 km) into a spatially explicit database and combined an air chemistry transport model and a health assessment model. By comparing the results with and without ETS, this analysis is intended to provide concrete evidence of the complex impacts of ETS on the environment and human health, thus bringing transparency to the ongoing discussion on the co-benefits of ETS on climate change mitigation and air pollution control.

The remainder of this paper was organized as follows. The methodologies and data were described in Section 8.2. Section 8.3 used a simulation analysis that compared carbon emissions and $PM_{2.5}$ concentrations with ETS and without ETS and estimated the number of premature deaths for each disease. Discussion and conclusions were presented in Sections 8.4 and 8.5, respectively.

8.2 Methods and data

8.2.1 Research framework

In this paper, we came up with an integrated research framework where the accounting of CO_2 and $PM_{2.5}$ emissions, pollutant transport modeling, and health impact assessment were undertaken successively. Specifically, we tabulated and compiled the inventories of CO_2 and various pollutants ($PM_{2.5}$, PM_{10}, SO_2, NO_x, etc.) between with and without ETS for all the corporations involved in Hubei's ETS. Then we used a transaction model to simulate the trading of carbon emission allowances between different corporations involved in Hubei's ETS for the year of 2015. The Community Multiscale Air Quality (CMAQ) model together with the Weather Research and Forecasting (WRF) model were combined as a basis for simulation of the variation in pollutant concentrations on different grid sites with and without ETS. Data on CO_2 emissions, pollutant concentrations, and population were further addressed through a geographic information system (GIS)-based spatial modeling platform. This was followed by the health assessment model to estimate the number of population affected by the resulting $PM_{2.5}$. By contrasting the situations between with and without the ETS, the complicated impacts of ETS on the environment and human health in Hubei can be explored in a comparative sense.

8.2.2 Accounting of CO_2 and $PM_{2.5}$ emissions

To understand the environmental and health impact of the ETS in Hubei, we accounted for the carbon and $PM_{2.5}$ emissions with ETS or without ETS, following the formulas:

$$E_{i,CO_2,1} = \sum M_{fuel,j} \times F_{fuel,j} + E_p \tag{8.1}$$

$$E_{i,CO_2,2} = E_{i,CO_2,1} \pm E_{trade} \tag{8.2}$$

where $E_{i,CO_2,1}$ is the CO_2 emissions of corporation i with ETS; M_{fuel} is the energy use of fuel type j; F_{fuel} is the CO_2 emission factor for a fuel type j; E_p is the CO_2 emissions from industrial processes; $E_{i,CO_2,2}$ is the carbon emissions of corporation i without ETS; E_{trade} is the trading v0lumes of ETS; "+" is for selling carbon quotas; and "–" is for buying quotas.

It was assumed that the air pollutant emissions produced by the enterprises were in proportional to CO_2 emissions despite the variations in end-of-pipe measures. We notice that such an assumption has also been made by other studies (Dong & Liang, 2014). The pollutant emissions of trading corporations with ETS could be obtained from the database directly, and those without ETS were calculated as follows:

$$P_{i,k,2} = \frac{P_{i,k,1} \times E_{i,CO_2,1}}{E_{i,CO_2,2}} \tag{8.3}$$

where $P_{i,k,2}$ is the emissions of pollutant k from corporation i without ETS; and $P_{i,k,1}$ is the emissions of pollutant k from corporation i with ETS.

8.2.3 The CMAQ model

The CMAQ model was developed by the US Environmental Protection Agency (Ching & Byun, 1999). We simulated the processes of pollutant conversion, diffusion, dry deposition flux, and wet deposition flux. The generated data comprised corporate-level emissions, urban household emissions, transport emissions and rural emissions caused by energy combustion. The changes in pollutant concentrations during the spatial transmission with and without ETS were determined on the basis of the CMAQ model. To obtain the $PM_{2.5}$ concentrations that people were exposed to, the inventory data of the trading corporations and the urban household and transport data were dealt with the CMAQ model on a 5 km × 5 km resolution. The resulting $PM_{2.5}$ concentrations in this paper were jointly determined by the primary pollution as well as the secondary transformation of pollutants such as SO_2 and NO_x, together with meteorological data, topographic parameters, etc. As a numerical weather prediction system designed to serve both atmospheric research and operational forecasting needs (Xie et al., 2018), the WRF model was developed by the US National Oceanic and Atmospheric Administration (NOAA) (Zhang et al., 2014).

8.2.4 Health impact assessment

We estimated the premature mortality with and without ETS by accounting for the fraction of cause-specific mortality attributable to long-term changes in $PM_{2.5}$ concentrations on the 5 km × 5 km resolution, following the methods by Silva et al. (2016). This health impact assessment consisted of several exposure-response functions (ERFs) based on global epidemiological studies (Burnett et al., 2004). Correspondingly, four representative diseases, namely ischemic heart disease (IHD), chronic obstructive pulmonary disease (COPD), stroke, and lung cancer

(LC), for which the risks of deaths mainly arise from the exposure to $PM_{2.5}$, were chosen to reflect the health impact due to air pollution. The ERFs of the above four diseases were uniform, with different coefficients (Table 8.1). The relative risk (RR) of deaths caused by $PM_{2.5}$ concentrations that people were exposed to was calculated as follows:

$$RR(C)_d = \begin{cases} 1+\partial_d\left(1-e^{-\gamma_d\left(C-C0_d\right)^{\delta_d}}\right), \text{ if } C > C0_d \\ 1, else \end{cases} \tag{8.4}$$

where C is the annual average of $PM_{2.5}$ concentration; $C0$ is the safe $PM_{2.5}$ concentration, below which there is no lethal probability of the above four diseases; α, γ, and δ are all coefficients used to describe the exposure-response curves; and d is the disease category. When the concentrations of $PM_{2.5}$ are lower than the threshold value of 10 μg/m³ (Cohen et al., 2017), RR is 1, indicating no health impacts. In this study, age groups were not taken into account due to data availability.

The attribute fraction (AF) serves as a statistical indicator that quantitatively describes the proportion of the mortality (or morbidity) caused by certain risk factors in the total mortality (or morbidity), following the equation below:

$$AF_d = \left(RR_d - 1\right)/RR_d \tag{8.5}$$

The number of premature deaths due to $PM_{2.5}$ concentrations is calculated as follows:

$$E_d = AF_d \times P \times B_d \tag{8.6}$$

$$AF_d = \left(RR_d - 1\right)/RR_d \tag{8.7}$$

where E is premature deaths; d is the type of diseases; P is population; and B is the baseline mortality rate for certain diseases, which was necessary to identify the exact number of grid sites where premature deaths occurred (Table 8.1).

Table 8.1 Exposure-response parameters and baseline mortality rates for different diseases

Diseases	α	γ	δ	*C0*	*Values*
IHD	0.83	0.0717	0.5516	6.96	17.61‰
Stroke	1.01	0.0174	1.1244	8.38	18.81‰
COPD	29.00	0.0005938	0.6786	7.17	9.13‰
LC	33.49	0.0000501	1.0128	7.24	6.08‰

Notes: Data were taken from the Hubei Provincial Center for Disease Control and Prevention. Abbreviations: IHD—ischemic heart disease; COPD—chronic obstructive pulmonary disease; LC—lung cancer.

8.2.5 Data

In 2015, there were 163 corporations included in ETS of Hubei which created a total emission trading volume of 10.72 million tons with a price range of 10.28-28.40 yuan per tons. There were 13 cities and four regions in Hubei involved in the ETS, including Ezhou, Enshi, Huanggang, Huangshi, Jingmen, Jingzhou, Qianjiang, Shennongjia, Shiyan, Suizhou, Tianmen, Wuhan, Xiantao, Xianning, Xiangyang, Xiaogan, and Yichang. As illustrated in Table 8.2, the CO_2 trading data were obtained from Hubei Carbon Emissions Trading Center. The CO_2 emission data were from the China High-Resolution Emission Database (CHRED). The $PM_{2.5}$ emission data for corporations exempted from the ETS were from the CHRED database and Multi-resolution Emission Inventory for China (MEIC) emission inventory. Real-time online monitoring data were from the Continuous Emission Monitoring Systems (CEMS) developed by the China National Environmental Monitoring Centre. To describe the ETS, we took the *2015 Hubei Carbon Emission Allowance Allocation Plan* issued by the Hubei Provincial Development and Reform Commission as a reference for the allocation of carbon emission allowances. The plan indicated that a baseline method was adopted for electricity, heat, and cement industries, whereas for other industries a historical method was utilized. The simulation period for both CO_2 and $PM_{2.5}$ ranged from January 2 to December 31, 2015. Besides, there were a total of 7 278 grid points in Hubei. We compared them with the monitoring concentrations taken from air quality sites in Hubei. Changes in the total population and the weight of the gridded in 2015 came from the LandScan database, which was developed by the Oak Ridge National Laboratory and has been widely used for spatial analysis due to the finest resolution of global population distribution (Zheng et al., 2019; Zhou et al., 2010). Moreover, we incorporated land use and terrain data from the National Aeronautics and Space Administration (NASA) into the WRF model to generate meteorological data. The meteorological

Table 8.2 Data sources of this study

Data	*Sources*
CO_2 emissions with ETS	China High-Resolution Emission Database (CHRED), Multi-resolution Emission Inventory for China (MEIC)
CO_2 emissions without ETS	Hubei Carbon Emissions Trading Center
Energy emission factors	National Development and Reform Commission (NDRC)
Exposure-response parameters and baseline mortality rates	Hubei Provincial Center for Disease Control and Prevention
Land use and terrain data	National Aeronautics and Space Administration (NASA)
Meteorological data	National Oceanic and Atmospheric Administration (NOAA), National Centers for Environmental Prediction (NCEP)
Pollutant emissions with ETS	China High Resolution Emission Database (CHRED), Multi-resolution Emission Inventory for China (MEIC)
Population	Landscan database
Real-time online monitoring data	China National Environmental Monitoring Centre

input data on a 1° grid were provided by the National Centers for Environmental Prediction (NCEP) (Woo et al., 2012).

8.3 Results

8.3.1 CO_2 emissions with and without ETS

In terms of the changes in CO_2 emissions from corporations with and without ETS, we find that more than 90% of the corporations had CO_2 emissions above 0.5 million tons without ETS, among which those exceeding 5 million tons were mainly distributed in Ezhou, Xiaogan, Huanggang, and especially in Wuhan. With the implementation of the ETS, 64 corporations experienced an increase in the CO_2 emissions. Amongst 11 corporations were found in Wuhan, in which three had CO_2 emissions above 3.5 million tons, and eight in Xiaogan, of which six exceeded 0.5 million tons and one exceeded 5 million tons. Reduced CO_2 emissions were observed in 101 corporations, whereas only one corporation remained unchanged under the ETS.

Correspondingly, more than 80% of the growth took place in the old industrial area of Wuhan with lots of carbon-intensive corporations and dense population. We observed that heavy industries tended to purchase carbon emission allowances through ETS, many of which already had a considerably high amount of emissions before the ETS. A large number of heavy industrial corporations with increased carbon emissions were located in areas that had dense populations and high concentrations of co-pollutants. In contrast, corporations whose carbon emissions were reduced through the ETS were widely dispersed throughout Hubei. This suggested that carbon emission allowances may be sold by corporations with smaller carbon emissions and purchased by carbon-intensive ones.

8.3.2 Spatial distribution of $PM_{2.5}$ concentrations

The spatial distribution of $PM_{2.5}$ concentrations with and without ETS shows a decline trend in $PM_{2.5}$ concentrations under the ETS, from 30.4 $\mu g/m^3$ without ETS to 29.9 $\mu g/m^3$ with ETS. Nevertheless, generally, the distribution patterns of pollutants with ETS were consistent with those without ETS. Overall, the highest $PM_{2.5}$ concentrations were always found in the Wuhan metropolitan area including Wuhan, Ezhou, Yichang, and Xiaogan, ranging from 100 to 120 $\mu g/m^3$. Among them, the pollutant concentration in Wuhan was the highest, followed by Ezhou, Xiaogan, and Yichang. The $PM_{2.5}$ concentrations in other areas of Hubei were relatively low. Similarly, the number of high-concentration grid sites decreased under the ETS. As a result, the $PM_{2.5}$ concentrations in Wuhan, Ezhou, Huangshi, and Yichang decreased by up to 8 $\mu g/m^3$, but still tremendously exceeded the 35 $\mu g/m^3$ limit of the secondary standard of the National Ambient Air Quality Standards. Specifically, the Qingshan District in Wuhan was the region with the highest $PM_{2.5}$ concentrations in Hubei, ranging from 100 to 140 $\mu g/m^3$.

In terms of the changes to the spatial distribution of $PM_{2.5}$ concentrations with and without ETS, we notice that there were 80 of 163 corporations whose pollutant emissions increased under the ETS. Among them, 13 corporations were found to increase by over 1 million tons, mainly belonging to the steel, power energy, chemical, and cement industries. Remarkably, more than 80% of the total increase can be attributed to three corporations in the steel industry. In addition, power energy was the industry in which two companies experienced a emission growth with ETS, accounting for about 12% of the total increase in air pollutants. These corporations were mainly located at the Qingshan Industrial Park in Wuhan. Interestingly, approximately 90% of the total increase in $PM_{2.5}$ concentrations was observed for Wuhan, particularly in its old center with lots of coal-dependent heavy industries, such as steel and cement (Ding et al., 2016). In this sense, it would be not surprising that the premature mortality in the old center of Wuhan, where corporations bought higher carbon emission allowances, was much higher than that in areas where many corporations sold allowances. This was analogous to the situation in Ezhou, Jingzhou, and Yichang where chemical and power energy played an important role in urban industries.

8.3.3 $PM_{2.5}$-related premature mortality associated with ETS

The most striking impact of exposure to air pollutants was premature death. The premature mortality from $PM_{2.5}$ pollution in Hubei was over 8‰ without ETS, including about 51.3% for males and 48.7% for females (Figure 8.1). Specifically, IHD, LC, COPD, and stroke were responsible for about 32.7%, 11.9%, 3.3%, and 52.1% of the total deaths in Hubei without ETS, respectively. By contrast, the mortality in Hubei decreased by 5 000 when it comes to ETS. In that case, the mortality of IHD, LC, COPD, and stroke accounted for 32.7%, 11.8%, 26.7%, and 51.8% of the total deaths, respectively. The largest decrease in premature death was observed for stroke (58%), followed by IHD (26%), COPD (12%), and LC (4%). However, there was an unexpected increase of 1 500 premature deaths under ETS, accounting for 0.15‰ of the total population in Wuhan as of 2015. Moreover, while for all the four diseases the premature deaths of males were higher than those of females, there was only a tiny difference in deaths decline between the two genders. Approximately 0.14 million people benefitted from the ETS that led to a decline in urban $PM_{2.5}$ concentrations, accounting for 0.12‰ of the total population in Hubei as of 2015.

Many studies show that long-term exposure to high $PM_{2.5}$ pollution may induce higher mortality of respiratory, cardiovascular, and cerebrovascular diseases (Cesaroni et al., 2013). Figure 8.2 illustrates the total number of premature mortalities in each city of Hubei with ETS and without ETS. Cities with higher health risks were mostly in the Wuhan metropolitan area. No matter whether the ETS was launched or not, Wuhan, Shiyan and Jingzhou were always the top three in terms of premature deaths which accounted for 41.9% of total premature mortality in Hubei. Conversely, the bottom three comprised Shennongjia, Ezhou, and Qianjiang. With the implementation of ETS, all the cities exhibited a notable reduction of premature deaths, ranging from 12 (Shennongjia) to 1100 (Wuhan) cases. The premature mortality in all the cities decreased by over 1%, especially for Jingmen where a

(a)

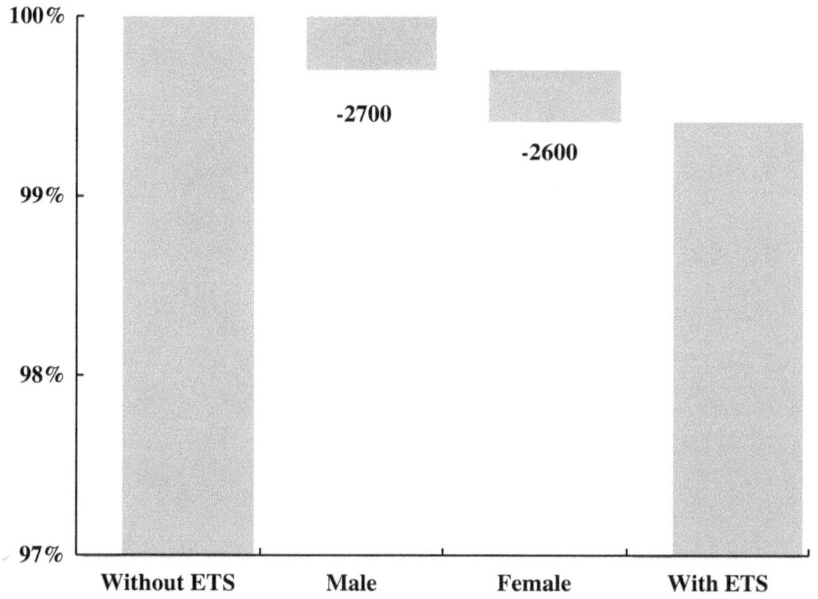

(b)

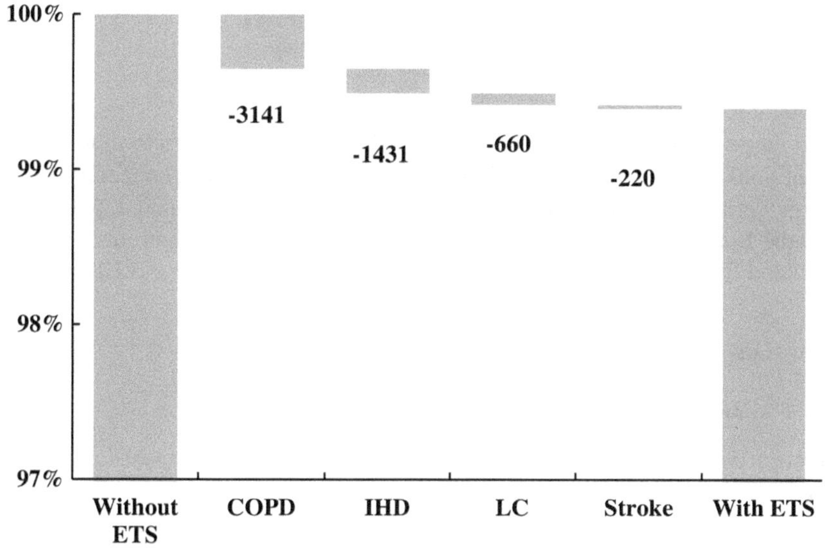

Figure 8.1 Mortality due to $PM_{2.5}$ pollution without and with ETS (*Continued*)

Note: (a) The total premature deaths by disease and by gender; (b) The decrease in premature deaths by gender; (c) The decrease in premature deaths by disease. Abbreviations: IHD—ischemic heart disease; COPD—chronic obstructive pulmonary disease; LC—lung cancer.

(c)

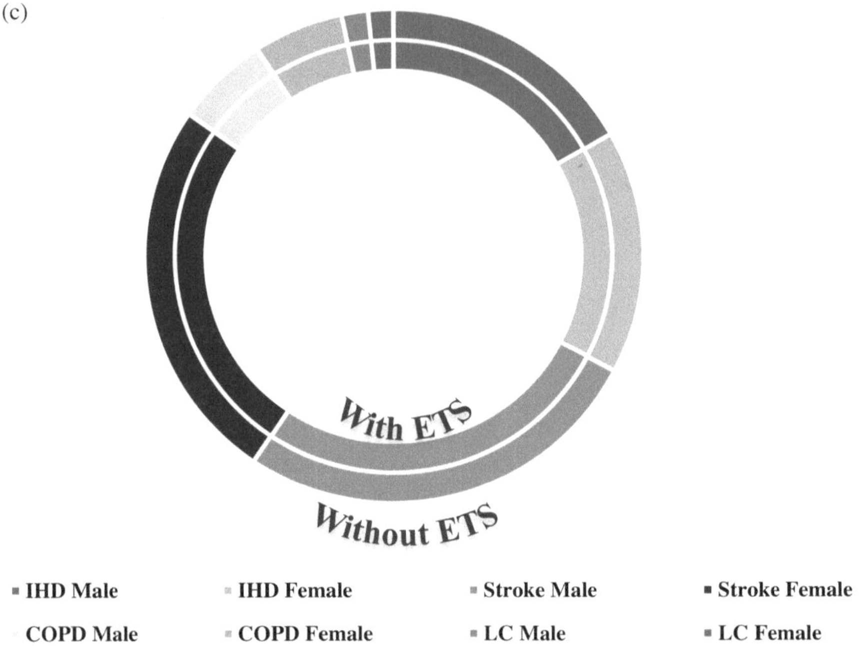

Figure 8.1 (Continued)

reduction of 1.5% was witnessed. However, because of the ETS, it became normal that corporations with smaller CO_2 emissions sold extra quotas to those with larger emissions. In some regions of Wuhan and Yichang, the $PM_{2.5}$ concentrations under ETS were higher than before, leading to an increase of around 1 500 premature deaths from these regions. At the grid level, Wuhan, Jingzhou, and Yichang accounted for over 99% of the total increase of premature deaths in Hubei.

8.4 Discussion

8.4.1 Comparison of simulated and monitoring data

To test the robustness of our analysis, data derived from the official air quality monitoring stations were employed to verify the air quality simulation model developed in this paper. A total number of 42 monitoring stations in Hubei were selected for comparison and verification of the simulation results. Considering that $PM_{2.5}$ are the main contributors to air pollution, Figure 8.3 presents both the simulated and monitoring values for the monthly $PM_{2.5}$ concentrations of four representative cities in 2015. One may conclude that the simulation undertaken in this paper was generally consistent with the monitoring data (with a relative error of mostly <0.2), and therefore provided a scientific basis for the research and the. It demonstrated that the simulated data could be used to our paper as the base data

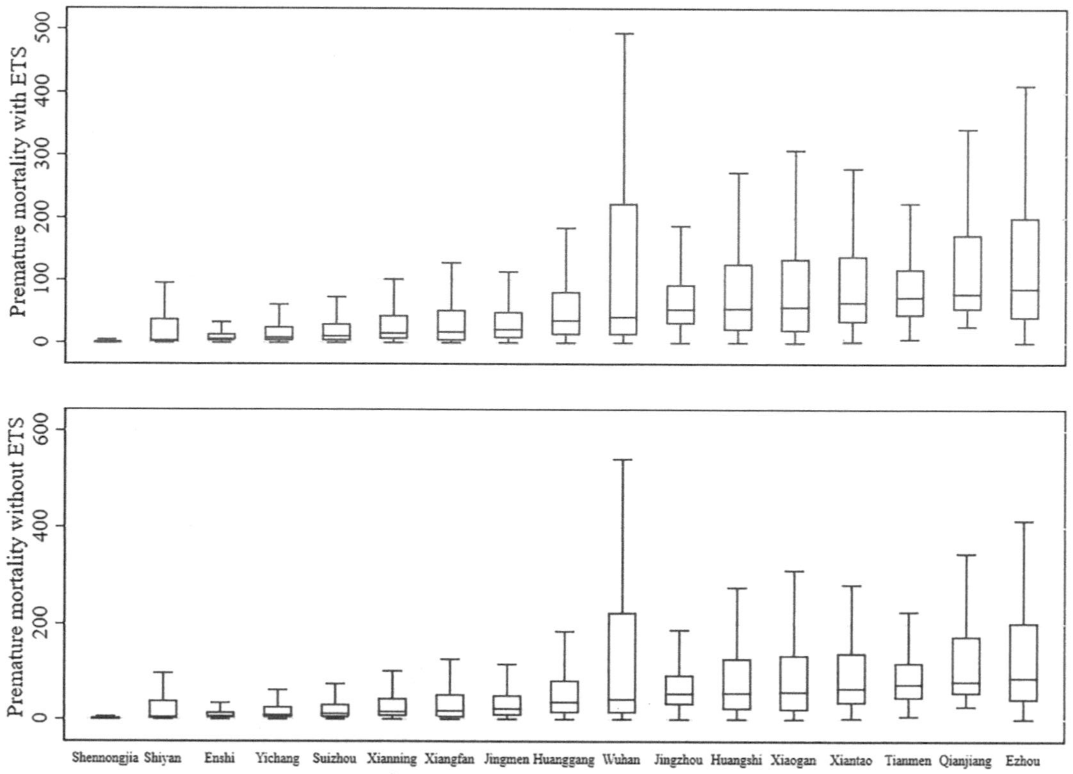

Figure 8.2 Premature mortality of 17 cities in Hubei without ETS and with ETS

Note: The cities were ordered according to the median number of premature deaths. The top line in the box diagram indicates the number of premature deaths on the grid sites with the most premature deaths in this city. The bottom line indicates the opposite. The three lines in the box from top to bottom are the upper quartile line (75%), the median line (50%), and the lower quartile line (25%).

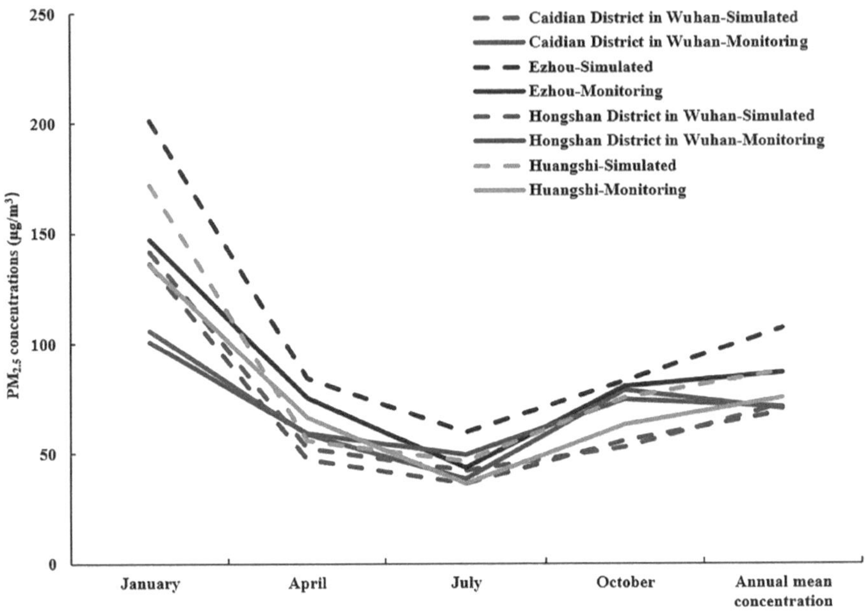

Figure 8.3 Comparison of simulated and monitoring $PM_{2.5}$ concentrations

to assess the impacts of health and environment and therefore provide a scientific basis for the research.

8.4.2 Reasons for the increase in $PM_{2.5}$ concentrations under ETS

The CO_2 emissions remained almost unchanged with ETS in Hubei. There were 7 278 grid sites in Hubei, 470 of which have an increase in $PM_{2.5}$ concentrations. Most of them were concentrated in Wuhan (122) and Yichang (63). For the latter, for instance, four corporations experienced an increase in $PM_{2.5}$ concentrations under the ETS. Two of them had $PM_{2.5}$ concentrations over 35 μg/m^3, and the remaining two were above 40 μg/m^3. Overall there were a majority of corporations in Hubei whose $PM_{2.5}$ concentrations decreased under the ETS. Because of the synergistic effect, the increase in $PM_{2.5}$ concentrations was often accompanied by the increase in CO_2 emissions on the same grid sites. Overall, fifteen corporations in Hubei were found to be on grid sites of this nature, 14 of which were buyers and merely one was a seller. For the latter, it was simply because that the allocated carbon emissions allowance was higher than its own emissions.

8.4.3 Uncertainty, limitations, and future work

Although this paper enabled an overall picture of the environmental and health impacts of ETS in Hubei on the basis of numerous corporate-level data, it still has some uncertainty and limitations that require further investigation. First, the variation in relative risk functions constituted a major source of uncertainty in $PM_{2.5}$-related mortality estimation (Burnett et al., 2014; Chen et al., 2017; San José et al., 2018). For example, China's $PM_{2.5}$-related premature deaths in 2013 were estimated to be 1.37 million by Liu et al. (2016) using the ERFs, whereas Fang et al. (2016) reported 3.03 million premature deaths using Late-life functions for the same year. As such, the premature mortality calculated based on the ERFs had a slightly higher value than that based on other functions and therefore may lead to an overestimation of $PM_{2.5}$-related deaths. Even though, our simulation made sense as the high spatiotemporal variation in $PM_{2.5}$ concentrations to which people were exposed is usually not captured by the fixed air quality monitoring stations (Menon & Nagendra, 2018). Second, the selection and integration of various models were far from satisfactory. Some of them may be not always compatible with each other. This is mainly due to the different applicability of all the methods/models. Thus, a full understanding of the limitations and appropriate use of each of them would be needed. Third, the benefits of ETS may be diverse and go beyond environmental sustainability and human health. For instance, resource efficiency and social welfare can be enhanced as a result of ETS (Zhang et al., 2017). However, data availability poses a barrier to evaluating the benefits of ETS of this nature. Moreover, in addition to the health impacts due to ambient air pollution, indoor air pollution has a non-negligible impact on human health as well (Wu et al., 2019). Last but not least, two ways in which the ETS is beneficial to air quality and human health can be identified. One is to reduce the total CO_2 emissions through restrictions on the emission allowances of Hubei. The other is to change the spatial distribution of emissions within the province. The other is to change the spatial distribution of emissions within the province. It is challenging to quantify the contribution of each way explicitly, unless cross-provincial ETS works well and all the data are available.

To sum up, the impact of China's ETS on the environment and human health remains to be fully investigated. Since China's nationwide ETS at a very early stage, there is a need for further clarity and transparency on some basic information regarding the ETS, such as trading volume and price, and the coverage of corporations and industries involved. To overcome the limitations mentioned above, further improvements need to focus on the development of a robust methodology as a basis for revealing the co-benefits and air pollution-related premature deaths associated with the ETS, as well as updates in the CHRED. Scenario analysis may have the potential to facilitate seeking for the pathways toward optimal ETS within and beyond Hubei.

8.5 Conclusions and implications

China has already announced to develop a nationwide ETS with the goals of energy savings and emissions reduction. To measure the environmental and health impacts of ETS, this paper came up with a unifying framework that incorporated the accounting of CO_2 and $PM_{2.5}$ emissions, pollutant transport modeling, and health impact assessment into the simulation analysis. In this empirical study, we concluded that the ETS improved air quality in large parts of Hubei, as proved by the notable decline in both $PM_{2.5}$ concentrations and the proportion of heavily polluted areas throughout Hubei. As a result, almost 0.14 million people benefitted from the ETS that led to lower $PM_{2.5}$ concentrations in most of the grid sites in Hubei, irrespective of the growing public health risk in several densely populated regions primarily in the Wuhan metropolitan area; and While the premature deaths of male for all the four diseases were higher than those of female, no sharp difference was observed in deaths decline between the two genders. Furthermore, the value-added of this research lies in: (1) Proposing a unifying framework that incorporated the accounting of CO_2 and $PM_{2.5}$ emissions, pollutant transport modeling, and health impact assessment into a high-resolution simulation analysis based on the corporate-level data derived from CHRED developed by the authors and other databases; (2) Providing concrete evidence of the complex impacts of ETS on the environment and human health across grid sites and regions; and (3) Serving as a reference for operationalizing the ETS as a basis for enhancing the co-benefits of carbon emissions reduction and air pollution control in China.

China is suffering from severe air pollution. In comparison to other climate policies, ETS generally has a relatively satisfactory performance on achieving carbon and pollutant emission reduction targets in a cost-effective manner (Chen et al., 2017). Given that population in high-emission areas are usually exposed to high $PM_{2.5}$ concentrations, ETS in China should be extended to multiple industries other than the power generation industry. However, the buying of carbon emission allowances or offset credits should be restrained to large emitters, such as key corporations located in areas with high atmospheric concentrations and dense populations. The ETS in California, USA, is one example. In this case, regions with increased pollutant concentrations are entitled to more trading quotas in order to reduce potential environmental inequality (Anderson et al., 2018; Cai et al., 2016). An alternative is the establishment of an "exchange rate" that makes the price for high-priority emitters higher than that for lower-priority emitters (Muller & Mendelsohn, 2009). Generally, policies related to the ETS should encourage polluters in these areas cutting emissions as rapidly as possible.

Atmospheric environmental monitoring should be seen as a critical step toward comprehensive market supervision of ETS, especially for areas where the concentration of pollutants increases due to ETS. Our study evidently showed a notable decrease in not only CO_2 emissions and $PM_{2.5}$ concentrations but also $PM_{2.5}$-related premature mortality for most parts of Hubei when implementing the ETS. It was found that nearly 0.14 million people in Hubei were exposed to lower $PM_{2.5}$ concentrations with the help of ETS, irrespective of the growing public health risk in some densely populated areas, most of which were in the Wuhan metropolitan area. This calls for collaborative actions of authorities and organizations to for sharing monitoring data, unifying emission standards, and improving market supervision for all the corporations in this area. In addition, labor-intensive industrial sectors, such as agriculture, food, textile, and service, were found to suffer more output losses from $PM_{2.5}$ pollution. It implies that corporations in these sectors should pay more attention to workers' health.

Furthermore, health impact should be an important criterion for evaluating the effects of ETS policies. As the industrial structure of Hubei has much in common with the national average, our findings could also be informative for making ETS policies in other parts of China. We argue that China would benefit substantially from implementing control measures under the ETS. To that end, health impact should be an important criterion for evaluating the effects of ETS policies, which however has somehow been neglected in present ETS studies. The government, producers, and consumers should work together to speed up transition toward a low-emission industrial system that could bring considerable environmental and health co-benefits (Lu et al., 2020). An effective ETS for this endeavor should give consideration to both fairness and efficiency in mitigating climate change and reducing air pollution, with a particular focus on central cities with dense populations.

References

Anderson, C. M., Kissel, K. A., Field, C. B., & Mach, K. J. (2018). Climate change mitigation, air pollution, and environmental justice in California. *Environmental Science & Technology*, *52*(18), 10829–10838.

Aunan, K., Fang, J., Hu, T., Seip, H. M., & Vennemo, H. (2006). Climate change and air quality-measures with co-benefits in China. *Environmental Science & Technology*, *40*, 4822–4829.

Bain, P. G., Milfont, T. L., Kashima, Y., Bilewicz, M., Doron, G., Garðarsdóttir, R. B., Gouveia, V. V., Guan, Y., Johansson, L. O., Carlota, P., Aragones, C. V., Utsugi, J. I., Demarque, A., Otto, C., Park, S., Soland, J., Steg, M., González, L., Lebedeva, R., Madsen, N., Wagner, O. J., Akotia, C., Kurz, C. S., Saiz, T., Schultz, J. L., Einarsdóttir, P. W., & Saviolidis, G. (2016). Co-benefits of addressing climate change can motivate action around the world. *Nature Climate Change*, *6*(2), 154–157.

Burnett, R. T., Pope, C. A. III, Ezzati, M., Olives, C., Lim, S. S., Mehta, S., Shin, H. H., Singh, G., Hubbell, B., Brauer, M., Anderson, H. R., Smith, K. R., Balmes, J. R., Bruce, N. G., Kan, H., Laden, F., Prüss-Ustün, A., Turner, M. C., Gapstur, S. M., Diver, W. R., & Aaron, C. (2014). An integrated risk function for estimating the global burden of disease attributable to ambient fine particulate matter exposure. *Environmental Health Perspectives*, *122*(4), 397–403.

Burnett, R. T., Stieb, D., Brook, J. R., Cakmak, S., Dales, R., Raizenne, M., Vincent, R., & Dann, T. (2004). Associations between short-term changes in nitrogen dioxide and mortality in Canadian cities. *Archives of Environmental Health: An International Journal*, *59*(5), 228–236.

Cai, B., Bo, X., Zhang, L., Boyce, J. K., Zhang, Y., & Lei, Y. (2016). Gearing carbon trading towards environmental co-benefits in China: Measurement model and policy implications. *Global Environmental Change*, *39*, 275–284.

Cesaroni, G., Badaloni, C., Gariazzo, C., Stafoggia, M., Sozzi, R., Davoli, M., & Forastiere, F. (2013). Long-term exposure to urban air pollution and mortality in a cohort of more than a million adults in Rome. *Environmental Health Perspective*, *121*(3), 324–331.

CHEE (China Hubei Emission Exchange). (2015). Carbon market conditions. http://www.tanpaifang.com/tanjiaoyi/2015/0403/43583.html.

Chen, L., Shi, M., Gao, S., Li, S., Mao, J., Zhang, H., Sun, Y., Bai, Z., & Wang, Z. (2017). Assessment of population exposure to $PM_{2.5}$ for mortality in China and its public health benefit based on BenMAP. *Environmental Pollution*, *221*, 311–317.

Cheng, B., Dai, H., Wang, P., Zhao, D., & Masui, T. (2015). Impacts of carbon trading scheme on air pollutant emissions in guangdong province of China. *Energy for Sustainable Development*, *27*, 174–185.

Ching, J., & Byun, D. (1999). Introduction to the models-3 framework and the community multiscale air quality model (CMAQ). Science Algorithms of the EPA Models-3 Community Multiscale Air Quality (CMAQ) Modeling System.

Cohen, A. J., Brauer, M., Burnett, R., Anderson, H. R., Frostad, J., Estep, K., Balakrishnan, K., Brunekreef, B., Dandona, L., Dandona, R., Feigin, V., Freedman, G., Hubbell, B., Jobling, A., Kan, H., Knibbs, L., Liu, Y., Martin, R., & Forouzanfar, M. H. (2017). Estimates and 25-year trends of the global burden of disease attributable to ambient air pollution: An analysis of data from the global burden of diseases study 2015. *The Lancet*, *389*(10082), 1907–1918.

Cushing, L., Blaustein-Rejto, D., Wander, M., Pastor, M., Sadd, J., Zhu, A., & Morello-Frosch, R. (2018). Carbon trading, co-pollutants, and environmental equity: Evidence from California's cap-and-trade program (2011–2015). *PLOS Medicine*, *15*(7), e1002604.

Ding, L., Liu, C., Huang, Y., Chen, K., & Zeng, K. (2016). Spatial and temporal characteristics of urban ambient air quality and its main influence factors in Hubei province. *Economic Geography*, *36*(3), 170–178.

Dong, H., Dai, H., Dong, L., Fujita, T., Geng, Y., Klimont, Z., Inoue, T., Bunya, S., Fujii, M., & Masui, T. (2015). Pursuing air pollutant co-benefits of CO_2 mitigation in China: A provincial leveled analysis. *Applied Energy*, *144*, 165–174.

Dong, L., & Liang, H. (2014). Spatial analysis on China's regional air pollutants and CO_2 emissions: Emission pattern and regional disparity. *Atmospheric Environment*, *92*, 280–291.

Fang, D., Chen, B., Hubacek, K., Ni, R., Chen, L., Feng, K., & Lin, J. (2019). Clean air for some: Unintended spillover effects of regional air pollution policies. *Science Advances*, *5*(4), eaav4707.

Fang, K., Tang, Y., Zhang, Q., Song, J., Wen, Q., Sun, H., Ji, C., & Xu, A. (2019). Will China peak its energy-related carbon emissions by 2030? Lessons from 30 Chinese provinces. *Applied Energy*, *255*, 113852.

Fang, K., Wang, T., He, J., Wang, T., Xie, X., Tang, Y., Shen, Y., & Xu, A. (2020). The distribution and drivers of $PM_{2.5}$ in a rapidly urbanizing region: The Belt and Road Initiative in focus. *Science of the Total Environment*, *716*, 137010.

Fang, D., Wang, Q. G., Li, H., Yu, Y., Lu, Y., & Qian, X. (2016). Mortality effects assessment of ambient $PM_{2.5}$ pollution in the 74 leading cities of China. *Science of the Total Environment*, *569*, 1545–1552.

GEA (Global Energy Assessment). (2012). Global energy assessment-toward a sustainable future. Cambridge, UK and New York, NY, USA, and the International Institute for Applied Systems Analysis. Cambridge University Press.

Haines, A., Amann, M., Borgford-Parnell, N., Leonard, S., Kuylenstierna, J., & Shindell, D. (2017). Short-lived climate pollutant mitigation and the sustainable development goals. *Nature Climate Change*, *7*(12), 863–869.

Haines, A., McMichael, A. J., Smith, K. R., Roberts, I., Woodcock, J., Markandya, A., Armstrong, B., Lendrum, D. C., Dangour, A. D., Davies, M., Bruce, N., Tonne, C., Barrett, M., & Wilkinson, P. (2009). Public health benefits of strategies to reduce greenhouse-gas emissions: Overview and implications for policy makers. *The Lancet*, *374*(9707), 2104–2114.

HDRC (Hubei Development and Reform Commission). (2017). Hubei Province Carbon Emission Allowance Allocation Plan in 2015.

IPCC (Intergovernmental Panel on Climate Change). (2017). FAO/IPCC Ecpert Meeting on land use, climate change and food security. https://www.ipcc.ch/publication/fao-ipcc-expert-meeting-on-land-use-climate-change-and-food-security/.

Lelieveld, J., Evans, J. S., Fnais, M., Giannadaki, D., & Pozzer, A. (2015). The contribution of outdoor air pollution sources to premature mortality on a global scale. *Nature*, *525*(7569), 367–371.

Li, L., Hong, X., & Wang, J. (2020). Evaluating the impact of clean energy consumption and factor allocation on China's air pollution: A spatial econometric approach. *Energy*, *195*, 116842.

Li, M., Zhang, D., Li, C. T., Mulvaney, K. M., Selin, N. E., & Karplus, V. J. (2018). Air quality co-benefits of carbon pricing in China. *Nature Climate Change*, *8*(5), 398–403.

Liu, Z., Guan, D., Wei, W., Davis, S. J., Ciais, P., Bai, J., Peng, S., Zhang, Q., Hubacek, K., Marland, G., Andres, R. J., Crawfor-Brown, D., Lin, J., Zhao, H., Hong, C., Boden, T. A., Feng, K., Peters, G. P., Xi, F., Liu, J., Li, Y., Zhao, Y., Zeng, N., & He, K. (2015). Reduced carbon emission estimates from fossil fuel combustion and cement production in China. *Nature*, *524*(7565), 335–338.

Liu, J., Han, Y., Tang, X., Zhu, J., & Zhu, T. (2016). Estimating adult mortality attributable to $PM_{2.5}$ exposure in China with assimilated $PM_{2.5}$ concentrations based on a ground monitoring network. *Science of the Total Environment*, *568*, 1253–1262.

Lu, Q., Fang, K., Heijungs, R., Feng, K., Li, J., Wen, Q., Li, Y., & Huang, X. (2020). Imbalance and drivers of carbon emissions embodied in trade along the Belt and Road Initiative. *Applied Energy*, *280*, 115934.

Menon, J. S., & Nagendra, S. S. (2018). Personal exposure to fine particulate matter concentrations in central business district of a tropical coastal city. *Journal of the Air & Waste Management Association*, *68*(5), 415–429.

Muller, N. Z., & Mendelsohn, R. (2009). Efficient pollution regulation: Getting the prices right. *American Economic Review*, *99*(5), 1714–1739.

Nemet, G. F., Holloway, T., & Meier, P. (2010). Implications of incorporating air-quality co-benefits into climate change policymaking. *Environmental Research Letters*, *5*, 014007.

Rafaj, P., Schöpp, W., Russ, P., Heyes, C., & Amann, M. (2013). Co-benefits of post-2012 global climate mitigation policies. *Mitigation and Adaptation Strategies for Global Change*, *18*(6), 801–824.

San José, R., Pérez, J. L., Pérez, L., & Barras, R. M. G. (2018). Effects of climate change on the health of citizens modelling urban weather and air pollution. *Energy*, *165*, 53–62.

Schmale, J., Shindell, D., Von Schneidemesser, E., Chabay, I., & Lawrence, M. (2014). Air pollution: Clean up our skies. *Nature*, *515*, 335–337.

Silva, R. A., West, J. J., Lamarque, J. F., Shindell, D. T., Collins, W. J., Dalsoren, S., Faluvegi, G., Folberth, G., Horowitz, L. W., Nagashima, T., Naik, V., Rumbold, S. T., Sudo, K., Takemura, T., Bergmann, D., Cameron-Smith, P., Cionni, I., Doherty, R. M., Eyring, V., Josse, B., MacKenzie, I. A., Plummer, D., Righi, M., Stevenson, D. S., Strode, S., Szopa, S., & Zeng, G. (2016). The effect of future ambient air pollution on human premature mortality to 2100 using output from the ACCMIP model ensemble. *Atmospheric Chemistry and Physics*, *16*(15), 9847–9862.

Silva, R. A., West, J. J., Lamarque, J. F., Shindell, D. T., Collins, W. J., Faluvegi, G., Folbeth, G. A., Horowitz, L. W., Nagashima, T., Naik, V., Rumbold, S. T., Sudo, K., Takemura, T., Bergmann, D., Smith, P. C., Doherty, R. M., Josse, B., Mackenzie, I. A., Stevenson, D. S., & Zeng, G. (2017). Future global mortality from changes in air pollution attributable to climate change. *Nature Climate Change*, *7*(9), 647–651.

Sun, R., Kuang, D., & Chang, D. (2014). Analysis on carbon trading effects upon energy economic-environment and calculation of reasonable carbon price intervals. *China Population Resource & Environment*, *24*, 82–90.

The State Council of China. (2013). Air Pollution Prevention and Control Action Plan. http://www.gov.cn/jrzg/2013-09/12/content_2486918.htm.

Thompson, T. M., Rausch, S., Saari, R. K., & Selin, N. E. (2014). A systems approach to evaluating the air quality co-benefits of US carbon policies. *Nature Climate Change*, *4*(10), 917–923.

Vandyck, T., Keramidas, K., Kitous, A., Spadaro, J. V., Van Dingenen, R., Holland, M., & Saveyn, B. (2018). Air quality co-benefits for human health and agriculture counterbalance costs to meet Paris Agreement pledges. *Nature Communications*, *9*(1), 1–11.

Wang, S., Zhou, C., Wang, Z., Feng, K., & Hubacek, K. (2017). The characteristics and drivers of fine particulate matter ($PM_{2.5}$) distribution in China. *Journal of Cleaner Production*, *142*, 1800–1809.

West, J. J., Smith, S. J., Silva, R. A., Naik, V., Zhang, Y., Adelman, Z., Fry, M. M., Anenberg, S., Horowitz, L. W., & Lamarque, J. F. (2013). Co-benefits of mitigating global greenhouse gas emissions for future air quality and human health. *Nature Climate Change*, *3*(10), 885–889.

Woo, J. H., Choi, K. C., Kim, H. K., Baek, B. H., Jang, M., Eum, J. H., Song, C. H., Ma, Y., Sunwoo, Y., Chang, L., & Yoo, S. H. (2012). Development of an anthropogenic emissions processing system for Asia using SMOKE. *Atmospheric Environment*, *58*, 5–13.

Wood, R., Grubb, M., Anger-Kraavi, A., Pollitt, H., Rizzo, B., Alexandri, E., Stadler, K., Moran, D., Hertwich, E., & Tukker, A. (2020). Beyond peak emission transfers: Historical impacts of globalization and future impacts of climate policies on international emission transfers. *Climate Policy*, *20*(S1), S14–S27.

Wu, S., Zheng, X., You, C., & Wei, C. (2019). Household energy consumption in rural China: Historical development, present pattern and policy implication. *Journal of Cleaner Production*, *211*, 981–991.

Xie, Y., Dai, H., Xu, X., Fujimori, S., Hasegawa, T., Yi, K., Masui, T., & Kurata, G. (2018). Co-benefits of climate mitigation on air quality and human health in Asian countries. *Environment International*, *119*, 309–318.

Yu, F., Han, F., & Cui, Z. (2015). Reducing carbon emissions through industrial symbiosis: A case study of a large enterprise group in China. *Journal of Cleaner Production*, *103*, 811–818.

Zeng, Y., Weishaar, S. E., & Vedder, H. H. (2018). Electricity regulation in the Chinese national emissions trading scheme (ETS): Lessons for carbon leakage and linkage with the EU ETS. *Climate Policy*, *18*(10), 1246–1259.

Zhang, H., Chen, G., Hu, J., Chen, S. H., Wiedinmyer, C., Kleeman, M., & Ying, Q. (2014). Evaluation of a seven-year air quality simulation using the weather research and forecasting (WRF)/Community multiscale air quality (CMAQ) models in the Eastern United States. *Science of the Total Environment*, *473*, 275–285.

Zhang, Q., & Fang, K. (2019). Comment on "Consumption-based versus production-based accounting of CO_2 emissions: Is there evidence for carbon leakage?". *Environmental Science and Policy*, *101*, 94–96.

Zhang, X., Zhang, X., & Chen, X. (2017). Valuing air quality using happiness data: The case of China. *Ecological Economics*, *137*, 29–36.

Zhang, D., Zhang, Q., Qi, S., Huang, J., Karplus, V. J., & Zhang, X. (2019). Integrity of firms' emissions reporting in China's early carbon markets. *Nature Climate Change*, *9*(2), 164–169.

Zheng, H., Zhao, B., Wang, S., Wang, T., Ding, D., Chang, X., Liu, K., Xing, J., Dong, Z., Aunan, K., Liu, T., Wu, X., Zhang, S., & Wu, Y. (2019). Transition in source contributions of $PM_{2.5}$ exposure and associated premature mortality in China during 2005–2015. *Environment International*, *132*, 105111.

Zhou, Y., Fu, J. S., Zhuang, G., & Levy, J. I. (2010). Risk-based prioritization among air pollution control strategies in the Yangtze River Delta, China. *Environmental Health Perspectives*, *118*(9), 1204–1210.

Appendix A

Table A.1 refers to the aggregation of sectors into five and three domains, and Table A.2 investigates the energy data for calculating the carbon intensity altogether with the GDP of corresponding years.

Table A.1 Sectors integration for IoT

Agriculture, Hunting, Forestry, and Fishing	Agriculture and forestry	Agriculture
Water Conservancy, Environment, and Public Facilities Management		
Mining and Quarrying	Industry	Industry
Oil and Gas Products		
Metal and Mining Products		
Non-metal and Other Mining Products		
Food and Tobacco		
Textile Products		
Leather and Footwear, Wood and Products of Wood and Cork		
Wood products and furniture		
Paper printing and stationery		
Coke, Refined Petroleum, and Nuclear Fuel		
Chemicals and Chemical Products		
Rubber and Plastics, Other Non-Metallic Mineral		
Basic Metals and Fabricated Metal		
Metal products		
Universal machinery		
Professional machinery		
Transport Equipment		
Electrical and Optical Equipment		
Communications Equipment, Computers, and Other Electronic Equipment		
Instruments and Apparatuses		
Other Machinery Products		
Scarp and Waste		
Metal Products, Machinery, and Equipment Repair Services		
Gas Supply		
Water Supply		
Electricity and Heat Supply		
Construction	Construction	

(*Continued*)

Table A.1 (Continued)

Transportation, Storage and Postal Service	Transportation, Storage and Postal Service	Service
Wholesale and retail	Wholesale and retail	
Hotels and Restaurants		
Information Transmission, Software, and IT Services		
Financial Intermediation		
Real Estate Activities		
Renting of M&Eq and Other Business Activities		
Scientific Research and Tech Service		
Other Community, Social, and Personal Services		
Education		
Health and Social Work		
Culture, Sports and Entertainment		
Public Admin and Defence; Compulsory Social Security		

Table A.2 Energy input of sectors

Sectors	*2002*		*2007*	*2012*									
Energy / *Years*	*Coal in total (mt)*	*Petroleum in total (mt)*	*Standard coal (mt)*	*Coal in total (mt)*	*Coke oven gas (bm³)*	*Heat (mkJ)*	*Electricity (bkWh)*	*Petrol (mt)*	*Kerosene (mt)*	*Diesel oil (mt)*	*Fuel oil (mt)*	*Petrol coke (mt)*	*Natural gas (bm³)*
Agriculture and forestry	2.32	0.05	2.91	1.16	0.07	0	0.39	0.001	0.00	0.002	0	0.06	0
Industry	12.67	0.20	50.26	28.40	0.67	5,261,300	65.48	0.05	0.01	0.20	0.13	0.41	0.37
Construction	0.04	0.04	0.52	0.20	0.005	0	1.03	0.05	0	0.11	0	0	0
Transportation, storage, and postal Service	0.09	0.86	3.13	0.15	0.001	0	2.59	1.09	0.10	2.22	0.02	0	0
Wholesale and retail	1.42	0.10	4.34	10.74	0.18	0	4.14	0.21	0.00	0.46	0	0	0

Table A.3 Material coefficient for key materials

Materials	*Value of coefficient*
Tailings from iron ore (t/t)	1.74
Iron ore for steel (t/t)	1.6
Limestone for cement (t/t)	1.15
Tailings from non-metal ore (t/t)	0.4

Table A.4 Main industrial products and production

Industries	*Representative company*	*Main product*	*Production volume*
Aluminum industry	China national aluminum company	Alumina Aluminum	1.2 Mt/year 0.43 Mt/year
Phosphorus chemical	Guizhou Kailin company; Kaiyang phosphorus-coal chemical industrial base; Xifeng phosphorus-coal chemical industrial base; WQingzhen phosphorus-coal chemical industrial base	Phosphorus ore Phosphorus Phosphorus fertilizer	10.05 Mt/year 78 kt/year 1.65 Mt/year
Iron/steel industry	Guiyang iron/steel company	Ferroalloy Steel	600 kt/year 600 kt/year
Coal chemical	Kaiyang Phosphorus-coal chemical industrial base; Xifeng phosphorus-coal chemical industrial base; WQingzhen phosphorus-coal chemical industrial base	Coke Ammonia	1.14 Mt/year 477 kt/year

Appendix B

Table B.1 Abbreviations of industrial subsectors investigated in this study

No.	*Abbreviation*	*Full name*	*Sector category*
1	CMD	Coal mining and dressing	*Mining*
2	PNGE	Petroleum and natural gas extraction	*Mining*
3	FMMD	Ferrous metals mining and dressing	*Mining*
4	NMMD	Nonferrous metals mining and dressing	*Mining*
5	NMD	Nonmetal minerals mining and dressing	*Mining*
6	ANFP	Agricultural nonstaple food processing	*Manufacturing*
7	FP	Food production	*Manufacturing*
8	BP	Beverage production	*Manufacturing*
9	TP	Tobacco processing	*Manufacturing*
10	TI	Textile industry	*Manufacturing*
11	GOFF	Garments and other fiber products	*Manufacturing*
12	LFDRP	Leather, furs, down, and related products	*Manufacturing*
13	TPP	Timber processing, bamboo, cane, palm fiber, and straw products	*Manufacturing*
14	FM	Furniture manufacturing	*Manufacturing*
15	PPP	Papermaking and paper products	*Manufacturing*
16	MR	Printing and record medium reproduction	*Manufacturing*
17	CESA	Cultural, educational, and sports articles	*Manufacturing*
18	PNFPC	Petroleum and nuclear fuel processing and coking	*Manufacturing*
19	RCMP	Raw chemical materials and chemical products	*Manufacturing*
20	MPP	Medical and pharmaceutical products	*Manufacturing*
21	CFP	Chemical fiber products	*Manufacturing*
22	RPP	Rubber and plastic products	*Manufacturing*
23	NMP	Nonmetal mineral products	*Manufacturing*
24	SPFM	Smelting and pressing of ferrous metals	*Manufacturing*
25	SPNM	Smelting and pressing of nonferrous metals	*Manufacturing*
26	MP	Metal products	*Manufacturing*
27	GPM	General-purpose machinery	*Manufacturing*
28	M	Special purpose machinery	*Manufacturing*
29	TE	Transportation equipment	*Manufacturing*
30	EME	Electrical machinery and equipment	*Manufacturing*
31	TREE	Telecommunication and other electrical equipment	*Manufacturing*
32	MIM	Measuring instrument machinery	*Manufacturing*
33	PHP	Production and distribution of electric and heat power	*PDEHGW*
34	PDF	Production and distribution of gas	*PDEHGW*
35	PDW	Production and distribution of water	*PDEHGW*
36	OI	Other industries (including support mining activities, artwork products, recycling and disposal of waste, and repair service of equipment)	*Manufacturing*

Appendix C

Table C.1 The ranking of sectors by carbon intensity

Code	*Sector (Abbreviation)*	*Code*	*Sector (Abbreviation)*
1	Electricity and hot water production and supply (*E&W Prod*)	16	Metal products (*MetlProd*)
2	Metallurgy (*Metl*)	17	Wholesale and retailing (*Whol&Ret*)
3	Nonmetal products (*NonmetlProd*)	18	Specialist machinery (*SpecMac*)
4	Transport equipment (*TranEqp*)	19	Wood processing and furnishing (*Wood&Furn*)
5	Petroleum refining, coking, etc. (*Petr&Coking*)	20	Construction (*Constr*)
6	Chemical industry (*ChemInd*)	21	Transport equipment (*TranEqp*)
7	Coal mining (*CoalMin*)	22	Clothing, leather, fur, etc. (*Clothing*)
8	Petroleum and gas (*Petr&Gas*)	23	Other services (*OthrServ*)
9	Paper making, printing, stationery, etc. (*Paper*)	24	Gas production and supply (*GasProd*)
10	Nonmetal mining (*NonmetlMin*)	25	Water production and supply (*WatProd*)
11	Textile (*Textile*)	26	Electrical equipment (*ElecEqp*)
12	Food processing and tobaccos (*Food&Tob*)	27	Instrument and meter (*Inst&Metr*)
13	Metal mining (*MetlMin*)	28	Electronic equipment (*ElenEqp*)
14	General machinery (*GenMac*)	29	Other manufacturing (*OthrMan*)
15	Agriculture (*Agr*)		

Table C.2 The ranking of sectors by trade intensity

Code	*Sector (Abbreviation)*	*Code*	*Sector (Abbreviation)*
1	Clothing, leather, fur, etc. (*Clothing*)	16	Gas production and supply (*GasProd*)
2	Electronic equipment (*ElenEqp*)	17	Food processing and tobaccos (*Food&Tob*)
3	Metal products (*MetlProd*)	18	Nonmetal products (*NonmetlProd*)
4	Electricity and hot water production and supply (*E&W Prod*)	19	Transport equipment (*TranEqp*)
5	Instrument and meter (*Inst&Metr*)	20	Metal mining (*MetlMin*)
6	Textile (Textile)	21	Other services (*OthrServ*)
7	Agriculture (*Agr*)	22	Electrical equipment (*ElecEqp*)
8	General machinery (*GenMac*)	23	Nonmetal mining (*NonmetlMin*)
9	Paper making, printing, stationery, etc. (*Paper*)	24	Other manufacturing (*OthrMan*)
10	Specialist machinery (*SpecMac*)	25	Construction (*Constr*)
11	Petroleum refining, coking, etc. (*Petr&Coking*)	26	Petroleum and gas (*Petr&Gas*)
12	Coal mining (*CoalMin*)	27	Wholesale and retailing (*Whol&Ret*)
13	Chemical industry (*ChemInd*)	28	Wood processing and furnishing (*Wood&Furn*)
14	Metallurgy (*Metl*)	29	Water production and supply (*WatProd*)
15	Transport equipment (*TranEqp*)		

Index

Notes: *Italic* page numbers refer to figures and **Bold** page numbers refer to tables.